Microsoft SharePoint Server 2019 and SharePoint Hybrid Administration

Deploy, configure, and manage SharePoint on-premises and hybrid scenarios

Aaron Guilmette

BIRMINGHAM - MUMBAI

Microsoft SharePoint Server 2019 and SharePoint Hybrid Administration

Commissioning Editor: Pavan Ramchandani
Acquisition Editor: Pavan Ramchandani
Content Development Editor: Aamir Ahmed
Senior Editor: Hayden Edwards
Technical Editor: Deepesh Patel
Copy Editor: Safis Editing
Project Coordinator: Kinjal Bari
Proofreader: Safis Editing
Indexer: Manju Arasan
Production Designer: Aparna Bhagat

First published: October 2020

Production reference: 1211020

Published by Packt Publishing Ltd.
Livery Place
35 Livery Street
Birmingham
B3 2PB, UK.

ISBN 978-1-80056-373-5

www.packt.com

I'd like to thank my children, who continue to feign interest when I tell them about my book projects, as well as my friends, who are sick of hearing about them. I'd also like to thank my girlfriend, Christine, who has patiently waited for three years for me to finish writing a series of books. I promise, I will come out of the basement and we will get to go on a date again.

I'd like to give a special thanks to the Packt development team members (namely Aamir Ahmed, Hayden Edwards, and Divij Kotian) that have continued to put up with me through five books.

Finally, I'd like to thank Microsoft for giving us products that have driven workplace and personal productivity for 40 years, empowering every person on the planet to do more. They continue to inspire the "learn it all" mindset in all of us.

– Aaron Guilmette

`Packt.com`

Subscribe to our online digital library for full access to over 7,000 books and videos, as well as industry leading tools to help you plan your personal development and advance your career. For more information, please visit our website.

Why subscribe?

- Spend less time learning and more time coding with practical eBooks and Videos from over 4,000 industry professionals

- Improve your learning with Skill Plans built especially for you

- Get a free eBook or video every month

- Fully searchable for easy access to vital information

- Copy and paste, print, and bookmark content

Did you know that Packt offers eBook versions of every book published, with PDF and ePub files available? You can upgrade to the eBook version at `www.packt.com` and as a print book customer, you are entitled to a discount on the eBook copy. Get in touch with us at `customercare@packtpub.com` for more details.

At `www.packt.com`, you can also read a collection of free technical articles, sign up for a range of free newsletters, and receive exclusive discounts and offers on Packt books and eBooks.

Contributors

About the author

Aaron Guilmette, a Teams technical specialist at Microsoft, provides guidance and assistance to customers adopting the Microsoft 365 platform. He primarily focuses on collaboration technologies, including Microsoft SharePoint Online, Microsoft Exchange, and Microsoft Teams. He also works with identity and scripting solutions.

He has been involved with technology since 1998 and has provided consulting services for customers in the commercial, educational, and government sectors internationally. Aaron has also worked on technical certification exams and instructional design for Microsoft and other organizations.

Aaron lives in Detroit, Michigan, with his five kids. When he's not busy solving technical problems, writing, or running kids to events, he's likely making a pizza and binge-watching dog videos.

About the reviewer

Sal Rosales is a Microsoft premier field engineer with over a decade of experience in assisting Fortune 500 companies with their collaboration needs. Sal specializes in Microsoft SharePoint and the Microsoft Power Platform. Solving technology-focused business gaps allows his customers to spend more time doing what they love; their happiness is the driving force behind Sal's work. Sal regularly participates in user community groups in which he is a strong advocate for diversity and inclusion in the workplace. Outside of work, Sal enjoys spending time with his family, training for athletic events, and surfing.

Yura Lee is a Technical Specialist at Microsoft with Office 365 and Azure consulting background. Today, she focuses on guiding and assisting customers in understanding the full potential of Microsoft Teams. Yura works with the whole collaborative stack including Exchange Online, SharePoint Online, Teams, Flow, and security components like Intune and Microsoft Information Protection. Yura lives in New Jersey with her family. In warmer weather, she and her boyfriend are most often exploring Italian restaurants and seeking out great IPAs in the greater tri-state area. In the winter, she's staying in, hosting dinner for her friends.

Packt is searching for authors like you

If you're interested in becoming an author for Packt, please visit `authors.packtpub.com` and apply today. We have worked with thousands of developers and tech professionals, just like you, to help them share their insight with the global tech community. You can make a general application, apply for a specific hot topic that we are recruiting an author for, or submit your own idea.

Table of Contents

Preface

SharePoint Server is an on-premises collaboration and business productivity platform. As a content management and web services platform, it enables users to create, publish, and discover content and applications, integrate with business systems, and create products.

Microsoft SharePoint Server 2019 and SharePoint Hybrid Administration, as the title implies, is designed to help you in understanding the platform, its tools, and its configuration capabilities in a way that will help you successfully deploy and configure a SharePoint farm. The SharePoint system is very broad. This book will walk you through the following concepts:

- Designing a SharePoint Server environment
- Identity and authentication
- Understanding service and web applications
- Connecting to external data systems
- Migrating to SharePoint Online

This book will help you understand the basics of the SharePoint platform and its architecture and terminology, along with the identity and authentication mechanisms used to provide access to resources. It will also guide you step by step through basic and advanced configurations for SharePoint Server, OneDrive, SharePoint hybrid connectivity, the on-premises data gateway, and Business Connectivity Services.

By the end of this book, you'll be equipped to confidently and successfully deploy and administer the collaboration platform components.

Who this book is for

Microsoft SharePoint Server 2019 and SharePoint Hybrid Administration is targeted at entry-level SharePoint Server administrators who want to learn how to deploy and manage SharePoint farms, service applications, and connected data services.

What this book covers

Chapter 1, *Overview of SharePoint Server 2019*, starts by explaining the basics of SharePoint Server and how it forms the foundation of all of the workloads that will be discussed later.

Chapter 2, *Planning a SharePoint Farm*, helps you understand the basic SharePoint MinRole server configurations and the architecture components that are necessary to build a successful solution.

Chapter 3, *Managing and Maintaining a SharePoint Farm*, shows you how to configure the core service applications for SharePoint Server farms.

Chapter 4, *Implementing Authentication*, explains the different identity and authentication methods available for securing web applications.

Chapter 5, *Managing Site Collections*, gives you a thorough understanding of the classic and modern SharePoint Server site collection architectures and provides recommendations on Microsoft's best practices.

Chapter 6, *Configuring Business Connectivity Services*, will show you how to interact with external data systems with Business Connectivity Services.

Chapter 7, *Planning and Configuring Managed Metadata*, shows you how the Managed Metadata Service application works and how to use it to build a structured taxonomy.

Chapter 8, *Managing Search*, dives into how to configure the Search service application.

Chapter 9, *Exploring Office Service Applications*, works through deploying some of the less-common Office-based service applications for document conversion and rendering.

Chapter 10, *Overview of SharePoint Hybrid*, introduces the concepts around integrating an on-premises SharePoint farm with SharePoint Online.

Chapter 11, *Planning a Hybrid Configuration and Topology*, helps you relate business requirements to SharePoint Hybrid features and plan for integration.

Chapter 12, *Implementing Hybrid Teamwork Artifacts*, walks you step-by-step through the configuration of several SharePoint Hybrid configurations, including hybrid sites, OneDrive, and taxonomy.

Chapter 13, *Implementing a Hybrid Search Service Application*, covers the different types of hybrid search options and the configuration of cloud hybrid search.

Chapter 14, *Implementing a Data Gateway*, shows you how to connect SharePoint Server to the Microsoft 365 platform to allow cloud-based tools, including Power BI and Power Automate, to interact with SharePoint Server data.

Chapter 15, *Using Power Automate with a Data Gateway*, demonstrates utilizing on-premises data with the Microsoft Power Automate service.

Chapter 16, *Overview of the Migration Process*, provides an overview of migrating to SharePoint Online using different tools.

Chapter 17, *Migrating Data and Content*, walks you through preparing data and migrating content to SharePoint Online.

To get the most out of this book

The SharePoint Server platform is best experienced (from an end user perspective) with either a laptop or desktop computer running a modern operating system, such as Windows 10 or macOS X 10.12 or later. Additionally, a modern browser, such as Microsoft Internet Explorer 11, Microsoft Edge, or a current version of Chrome, Safari, or Firefox, is necessary for the interfaces to operate correctly. If you are configuring services with integrated Windows authentication, you'll need your client computer to be a member of the domain where you have configured SharePoint services.

From a server and administration perspective, to follow the configuration examples, you'll need several servers on which to install Active Directory and SharePoint Server components. You can download a trial version of SharePoint Server 2019 from https:// www.microsoft.com/en-us/download/details.aspx?id=57462.

You will need a Microsoft 365 tenant for hybrid and data gateway configurations. You can sign up for a Microsoft 365 trial tenant (no credit card required) at https://www. microsoft.com/en-us/microsoft-365/business/compare-more-office-365-for-business-plans. You can also sign up for an Azure trial from the Azure portal (https:// portal.azure.com), which will allow you to provision virtual infrastructure if you do not have physical hardware for configuring SharePoint.

Some examples require various PowerShell modules, such as the SharePoint Online Management Shell (https://www.microsoft.com/en-us/download/details.aspx?id= 35588) or the Microsoft Teams module (https://www.powershellgallery.com/packages/ MicrosoftTeams/1.0.3).

Conventions used

There are a number of text conventions used throughout this book.

`CodeInText`: Indicates code words in text, database table names, folder names, filenames, file extensions, pathnames, dummy URLs, user input, and Twitter handles. Here is an example: "Modern site group membership is managed through the `Add-UnifiedGroupLinks` cmdlet."

A block of code is set as follows:

```
Set-SPSite -Site https://server.domain.com
```

When we wish to draw your attention to a particular part of a code block, the relevant lines or items are set in bold:

```
Set-SPUser -Site https://server.domain.com/sites/site1 -LoginName
<user@domain.com> -IsSiteCollectionAdmin $true
```

Any command-line input or output is written as follows:

```
PS C:>Connect-SPOService -Credential (Get-Credential) -Url
https://<tenant>-admin.sharepoint.com

Connected to SharePoint Online.
```

Bold: Indicates a new term, an important word, or words that you see onscreen. For example, words in menus or dialog boxes appear in the text like this. Here is an example: "In the navigation pane, select **Classic features**."

 Warnings or important notes appear like this.

 Tips and tricks appear like this.

Get in touch

Feedback from our readers is always welcome.

General feedback: If you have questions about any aspect of this book, mention the book title in the subject of your message and email us at customercare@packtpub.com.

Errata: Although we have taken every care to ensure the accuracy of our content, mistakes do happen. If you have found a mistake in this book, we would be grateful if you would report this to us. Please visit www.packtpub.com/support/errata, selecting your book, clicking on the Errata Submission Form link, and entering the details.

Piracy: If you come across any illegal copies of our works in any form on the Internet, we would be grateful if you would provide us with the location address or website name. Please contact us at copyright@packt.com with a link to the material.

If you are interested in becoming an author: If there is a topic that you have expertise in and you are interested in either writing or contributing to a book, please visit authors.packtpub.com.

Reviews

Please leave a review. Once you have read and used this book, why not leave a review on the site that you purchased it from? Potential readers can then see and use your unbiased opinion to make purchase decisions, we at Packt can understand what you think about our products, and our authors can see your feedback on their book. Thank you!

For more information about Packt, please visit packt.com.

Overview of SharePoint Server 2019

SharePoint Server 2019 is the latest Microsoft platform for rich document and workflow-based collaboration. Originally released as SharePoint Portal Server in 2001, it's continued to evolve over almost 20 years to become one of the premier document management and collaboration platforms for enterprises. It's been designed to help you to create, store, organize, visualize, and share data, content, and applications. Not only is SharePoint Server 2019 a content management platform, but it is also an application development platform. One of SharePoint Server's strengths is the ability to search for data across an enterprise content set, including connected resources such as other on-premises SharePoint Server environments or file servers.

In addition to content management capabilities for your documents and data, SharePoint Server can also connect to database services on-premises to produce data-driven dashboard pages. With SharePoint Hybrid, SharePoint Server can allow users to interact with data services and visualizations in Office 365 cloud-based solutions, and navigate between SharePoint Server and SharePoint Online environments seamlessly.

For content creators, SharePoint Server features an easy-to-use *what you see is what you get* drag and drop design, publishing, and categorization tools. For more precise design and advanced control of the end user experience, SharePoint Server also offers development interfaces for coding more complex sites and queries. As the owner of a SharePoint site, you can publish news, upload documents, add links to resources, and visualize data from other parts of the Office ecosystem or application widgets. Microsoft enables content consumers to access resources via browsers on desktop, tablet, and mobile devices, as well as native mobile applications for the Android and iOS platforms.

SharePoint Server features similar responsive design templates on its corresponding cloud service, SharePoint Online, as you can see in the following screenshot:

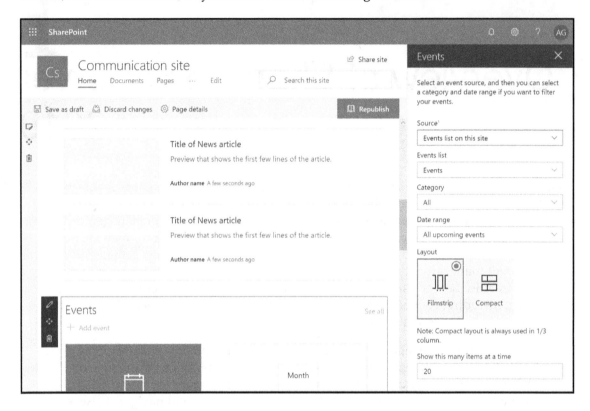

Responsive sites feature designs that are optimized for display on a variety of screens and devices, typically using a series of grids and flexible layouts (`https://docs.microsoft.com/en-us/office365/servicedescriptions/sharepoint-online-service-description/sharepoint-online-service-description#feature-availability-across-office-365-plans`).

While designing and managing responsive sites is not in the scope of this book, you may wish to further your knowledge in this area if you will be designing a SharePoint solution. You can learn more about SharePoint Server responsive design elements at `https://docs.microsoft.com/en-us/sharepoint/dev/design/grid-and-responsive-design`.

While SharePoint Server is frequently used for hosting local intranet sites or as a development platform for business process automation, it can also be used to host public websites or collaboration extranets with partners. SharePoint's workflows and security model make it simple to configure and promote content approval and publishing for public sites and to automate business processes when employees, partners, vendors, or customers submit or modify documents.

We'll explore some of SharePoint's core features and capabilities in the following sections:

- Overview of SharePoint Server architecture
- Overview of core SharePoint features and services

Understanding these features and capabilities will help you design and implement SharePoint Server environments to meet the needs of your organization.

Overview of SharePoint Server architecture

The SharePoint platform comprises farms, servers, site collections, sites, web parts, and pages.

A **farm** is a logical group of **servers** that have the SharePoint products installed and are configured to work together. A **site** is a website that contains various SharePoint objects, such as pages, document libraries, or calendars. A **site collection** is a group of sites, normally organized by department, project, cross-functional group, or other business units. A **page** is an HTML web page. Pages can be basic, just displaying text, or built from templates (such as a wiki or a publishing portal) that may have different web parts preloaded.

Web parts are codeless widgets or apps that can be used to display or interact with information on a page. In the following diagram, **Managers (Site)** contains three web parts: a calendar, a task list, and a document library:

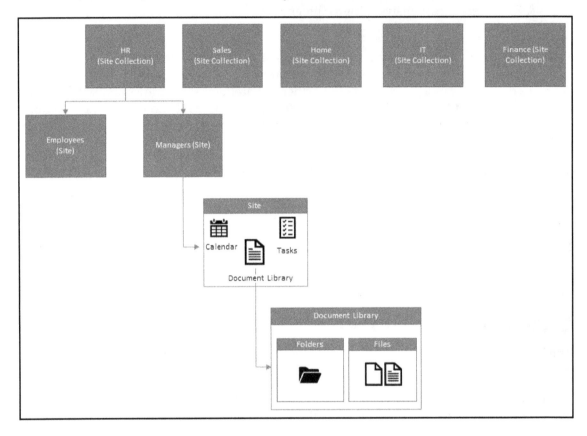

The **Document Library** web part looks like a filesystem interface to SharePoint and can be used to upload and download files, and to perform other file operation tasks.

In addition to the more visible components mentioned previously (such as servers, sites, and web parts), SharePoint Server has less visible components that are used to provide and control the features of the platform.

We'll look at these processes and services in the next section.

Overview of core SharePoint features and services

SharePoint is made up of a lot of moving pieces. While it's frequently used as an intranet or document collaboration workspace, it's also a robust application development platform, as mentioned earlier in this chapter. The SharePoint Server experience is fully customizable, and as a framework and platform for development, it is really only limited by the creativity of designers, developers, and administrators.

Many of SharePoint Server's features are made available through **service applications**. A service application can be thought of as a set of software and configuration components that make features and web applications available to users or other services. The following table lists the core SharePoint Server service applications and their functions:

Service application	Description
Access Services	Access Services in SharePoint Server provides service applications that enable you to share Access 2013 solutions on the web. Access Services has been deprecated for SharePoint Server 2019, but remains supported. For new design initiatives, Microsoft recommends using Power Apps and Power BI.
Business Data Connectivity Services	Business Data Connectivity Services allows you to connect to data stored and published in databases, web services, or an OData source for data operations.
Machine Translation Services	Machine Translation Services provides automated language conversion for publishing site collections.
Managed Metadata	The Managed Metadata service hosts a centrally shared term store, used for storing metadata (information about information) about the documents and assets in a SharePoint Server farm.
PerformancePoint Services	PerformancePoint Services is a performance management service that you can use to monitor and analyze your business through the use of dashboards, scorecards, and **key performance indicators (KPIs)**.

PowerPoint Automation Service	The PowerPoint Automation Service is used by SharePoint Server to perform conversions to PowerPoint documents.
SharePoint Search Service	The SharePoint Search Service is responsible for making content available to the search interfaces of SharePoint Server, allowing users to discover documents, applications, pages, and other data hosted in SharePoint.
Secure Store	The Secure Store Service is an authorization service that is used to store credentials.
State Service	The State Service is a shared service application used to store temporary data related to HTTP requests.
Subscription Settings and App Management Service	The Subscription Settings and App Management Service supports site subscriptions.
User Profile Service	The User Profile Service enables the creation and management of user profile data that can be used across SharePoint farms. The User Profile Service can synchronize data from Active Directory and make it available to farm applications.
Visio Services	SharePoint's Visio Services is used to render Visio diagrams in a web browser, letting users view diagrams without having to have Visio applications or custom viewers installed.
Word Automation Service	The Word Automation Service is used by SharePoint Server to perform conversions of documents supported by Microsoft Word.

Some service applications, such as Visio Services, provide a visible benefit to users (such as rendering Visio diagrams directly in a browser). Others, such as the State Service or User Profile Service, are used behind the scenes to make data available to other SharePoint applications.

When connected to Office 365 through on-premises data gateways, Office 365 Power Platform tools such as Power Automate (formerly Microsoft Flow), Power Apps, and Power BI can be used to interact with SharePoint Server data.

Summary

In this chapter, we introduced SharePoint Server, including the concepts of farms, servers, and site collections. You also learned how service applications are used to provide services to users and other services inside SharePoint Server. Finally, you learned about connecting to Power Platform applications in Office 365 using a data gateway.

The foundational knowledge of how farms, sites, and service applications work will be used in the upcoming chapters to show you how to design and implement successful SharePoint environments.

In the next chapter, we will begin learning about how to plan SharePoint farms, including the concepts of scalability and performance, as well as high availability and disaster recovery.

Planning a SharePoint Farm 2

SharePoint Server is deployed in a topology or layout called a **farm**. A farm is made up of one or more servers with specific roles or tasks assigned to them. Each of these roles performs different functions in the farm, and they work together to deliver a complete solution.

SharePoint Server deployments traditionally require architects and administrators to plan out the servers and services that are necessary. These architecture services typically include components for searching and indexing the database, the cache, and the application server roles.

While that fine-grained detail and control is still available, Microsoft has also developed a new deployment approach called MinRole. MinRole's design purpose is to optimize your system's resources and maximize performance while providing the best end user experience. To do this, administrators select from a set of roles to apply to a server when creating or joining a SharePoint Server farm.

As stated, the ability to manage SharePoint Server roles manually is still available. This book, however, will focus on deploying MinRole topologies. To learn more about custom roles, go to `https://docs.microsoft.com/en-us/sharepoint/install/planning-for-a-minrole-server-deployment-in-sharepoint-server`.

In this chapter, we'll focus on the necessary tasks to plan aspects of a SharePoint Server farm:

- Selecting and configuring a farm topology
- Designing for high availability
- Planning for disaster recovery
- Planning and configuring backup and restore options
- Planning for information rights management
- Planning localization and language packs
- Planning and configuring content farms
- Planning integration with Office 365 workloads
- Planning and configuring the OneDrive sync client
- Planning and configuring a high-performance farm

Mastering these planning tasks will allow you to successfully plan a robust SharePoint topology.

Selecting and configuring a farm topology

A SharePoint farm topology is used to describe the type or function of a particular SharePoint farm. When creating a SharePoint farm, roles are used to determine the functions of servers operating in a farm. There are eight server roles (seven predefined roles and one custom role), which are split across three categories:

- Dedicated roles
- Shared roles
- Special roles

These roles are used when deploying a chosen farm topology, as shown:

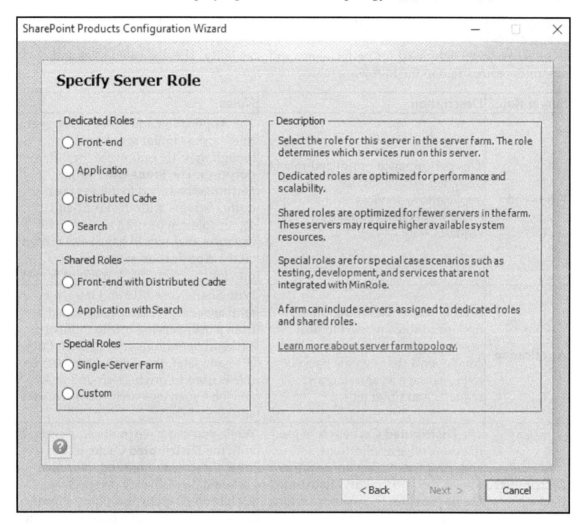

We'll look at those role categories and topologies next.

Dedicated roles

Dedicated roles are typically used in large-scale farms, but can also be used with smaller deployments or be added to servers that are configured for shared roles. The dedicated role layouts have been optimized for performance and scalability. The dedicated roles and descriptions are listed in the following table:

Server Role	Description	Notes
Front-end	This performance-optimized role is designed to host service applications, services, and components that render user requests.	The Application and Front-end server roles host a similar set of services, though they are optimized for different purposes. The **Front-end** role is performance-tuned to deliver user traffic. Servers with the **Front-end** server role can be used to run service instances that would have been hosted on the **Application** server role in previous versions of SharePoint Server.
Application	Servers with this throughput-optimized role are used to host service applications, services, and components that execute backend requests, such as search crawl requests and timer jobs.	With SharePoint 2016 and later, the term *application server* has changed from a user-serving role to more of a backend operations role. In SharePoint 2016 and later, the **Application** server role is used to run background tasks and jobs, such as search crawl requests and timer jobs.
Distributed Cache	The **Distributed Cache** role is used to host service applications, services, and components that are required for a distributed cache. **Distributed Cache** provides in-memory caching for features such as newsfeeds, authentication, and security trimming.	While you can have multiple servers with the **Distributed Cache** role in a farm, they are not redundant from a cached data resiliency perspective. Each server's cache is unique to itself. If a server hosting this role goes down, the data stored in its cache will be unavailable.
Search	The **Search** role is used to host service applications, services, and components that are required for search.	All servers with the **Search** role must be configured as part of the search topology before they can be used.

There may be situations where a farm design doesn't require dedicated roles, such as smaller departmental farms. In that case, you can use some of the pre-defined configurations, where functionalities are combined. We'll look at these **shared roles** next.

Shared roles

Shared role deployments are organized to allow a smaller number of servers in a farm by combining roles together. Shared role deployments can require more resources per server due to the increased number of components running. The shared roles and descriptions are listed in the following table:

Server Role	Description
Front-end with **Distributed Cache**	This is a shared role that combines the **Front-end** and **Distributed Cache** roles on the same server.
Application with **Search**	This is a shared role that combines the **Application** and **Search** roles on the same server.

Shared roles are common deployments for smaller farms (such as departmental farms or development farms). Next, we'll examine some special situations that require manual configuration.

Special roles

Finally, there is a special roles designation. Environments configured this way are typically used for development, testing, or for custom deployments where MinRole is not used to manage the services. The special roles and descriptions are listed in the following table:

Server Role	Description	Notes
Single-Server Farm	Servers configured with the **Single-Server Farm** role are appropriate for development, testing, or small-scale production tasks. The server will host all of the roles.	In previous versions of SharePoint, this was known as the standalone install mode. The **Single-Server Farm** role does have some differences than its predecessor—namely, the SharePoint administrator must configure the SQL server manually beforehand and must manually configure service applications afterward. A SharePoint farm using the **Single-Server Farm** role does not support more than one SharePoint server in the farm.
Custom	Servers configured with the **Custom** role will be managed manually.	This server role provides a lot of customization and flexibility, allowing the administrator to specifically determine which services will run. Typically, this role is used for applications or services that are not compatible or integrated with MinRole. MinRole cannot manage servers with the **Custom** role.

If you decide to later change a server's purpose on a farm, you can change its role using SharePoint Central Administration. The only stipulation is that you can only convert a server into a **Single-Server Farm** role if it's the only server in a farm.

In the next section, we'll review the services that will be configured with the various MinRole deployment options. The services installed with each role are important so that you can choose the appropriate server roles for your topology to ensure you have the right components, as well as optimal numbers of servers to support the high-availability and performance requirements.

Role services

The MinRole server configuration process installs and configures the appropriate services for each role. The following table lists the installed roles based on the MinRole option chosen. Services listed in *italics* are hidden as they are not internal to SharePoint and are not intended to be directly controlled:

Server Role	Installed Services
Front-end	Access Services Access Services 2010 App Management Service Business Data Connectivity Service Claims to Windows Token Service *Information Management Policy Configuration Service* Machine Translation Service The Managed Metadata web service *Microsoft Project Server Calculation Service* *Microsoft Project Server Events Service* *Microsoft Project Server Queuing Service* Microsoft SharePoint Foundation Administration Microsoft SharePoint Foundation Sandboxed Code Service Microsoft SharePoint Foundation Subscription Settings Service Microsoft SharePoint Foundation Timer *Microsoft SharePoint Foundation Tracing* *Microsoft SharePoint Foundation Usage* The Microsoft SharePoint Foundation web application Microsoft SharePoint Insights PerformancePoint Service The *Portal* service Project Server Application Service Request Management Secure Store Service *Security Token Service* *SSP Job Control Service* User Profile Service Visio Graphics Service

Server Role	Installed Services
Application	App Management Service Application Discovery and Load Balancer Service Business Data Connectivity Service Claims to Windows Token Service *Information Management Policy Configuration Service* Machine Translation Service The Managed Metadata web service *Microsoft Project Server Calculation Service* *Microsoft Project Server Events Service* *Microsoft Project Server Queuing Service* Microsoft SharePoint Foundation Administration Microsoft SharePoint Foundation Incoming E-Mail Microsoft SharePoint Foundation Subscription Settings Service Microsoft SharePoint Foundation Timer *Microsoft SharePoint Foundation Tracing* *Microsoft SharePoint Foundation Usage* The Microsoft SharePoint Foundation web application Microsoft SharePoint Foundation Workflow Timer Service Microsoft SharePoint Insights The *Portal* service The PowerPoint Conversion service Project Server Application Service Request Management Secure Store Service *Security Token Service* *SSP Job Control Service* User Profile Service Word Automation Services

Server Role	Installed Services
Distributed Cache	Claims to Windows Token Service (Prior to SharePoint Server 2016 FP 1) Distributed Cache Microsoft SharePoint Foundation Administration Microsoft SharePoint Foundation Timer Microsoft SharePoint Insights *Microsoft SharePoint Foundation Tracing* *Microsoft SharePoint Foundation Usage* The Microsoft SharePoint Foundation web application (Prior to SharePoint Server 2016 FP 1) The *Portal* service Request Management (Prior to SharePoint Server 2016 FP 1) *Security Token Service* *SSP Job Control Service*
Search	Application Discovery and Load Balancer Service Claims to Windows Token Service (Prior to SharePoint Server 2016 FP 1) Microsoft SharePoint Foundation Timer *Microsoft SharePoint Foundation Tracing* *Microsoft SharePoint Foundation Usage* Microsoft SharePoint Insights The *Portal* service Search Administration Web Service The Search Host Controller service Search Query and Site Settings Service *Security Token Service* SharePoint Server Search *SSP Job Control Service*
Custom	Distributed Cache Microsoft SharePoint Foundation Administration Microsoft SharePoint Foundation Timer The Microsoft SharePoint Foundation web application

Server Role	Installed Services
Single-Server Farm	Access Services
	Access Services 2010
	App Management Service
	Application Discovery and Load Balancer Service
	Business Data Connectivity Service
	Claims to Windows Token Service
	Distributed Cache
	Information Management Policy Configuration Service
	Lotus Notes Connector
	Machine Translation Service
	The Managed Metadata web service
	Microsoft Project Server Calculation Service
	Microsoft Project Server Events Service
	Microsoft Project Server Queuing Service
	Microsoft SharePoint Foundation Administration
	Microsoft SharePoint Foundation Incoming E-Mail
	Microsoft SharePoint Foundation Sandboxed Code Service
	Microsoft SharePoint Foundation Subscription Settings Service
	Microsoft SharePoint Foundation Timer
	Microsoft SharePoint Foundation Tracing
	Microsoft SharePoint Foundation Usage
	The Microsoft SharePoint Foundation web application
	Microsoft SharePoint Foundation Workflow Timer Service
	Microsoft SharePoint Insights
	PerformancePoint Service
	The *Portal* service
	PowerPoint Conversion Service
	Project Server Application Service
	Request Management
	Search Administration Web Service
	The Search Host Controller service
	Search Query and Site Settings Service
	Secure Store Service
	Security Token Service
	SharePoint Server Search
	SSP Job Control Service
	User Profile Service
	Visio Graphics Service
	Word Automation Services

Server Role	Installed Services
Front-end with **Distributed Cache**	Access Services Access Services 2010 App Management Service Business Data Connectivity Service Claims to Windows Token Service Distributed Cache *Information Management Policy Configuration Service* Machine Translation Service The Managed Metadata web service *Microsoft Project Server Calculation Service* *Microsoft Project Server Events Service* *Microsoft Project Server Queuing Service* Microsoft SharePoint Foundation Administration Microsoft SharePoint Foundation Sandboxed Code Service Microsoft SharePoint Foundation Subscription Settings Service Microsoft SharePoint Foundation Timer *Microsoft SharePoint Foundation Tracing* *Microsoft SharePoint Foundation Usage* The Microsoft SharePoint Foundation web application Microsoft SharePoint Insights PerformancePoint Service The *Portal* service Project Server Application Service Request Management Secure Store Service *Security Token Service* *SSP Job Control Service* User Profile Service Visio Graphics Service

Server Role	Installed Services
Application with **Search**	App Management Service Application Discovery and Load Balancer Service Business Data Connectivity Service Claims to Windows Token Service *Information Management Policy Configuration Service* Machine Translation Service The Managed Metadata web service Microsoft Project Server Calculation Service Microsoft Project Server Events Service Microsoft Project Server Queuing Service Microsoft SharePoint Foundation Administration Microsoft SharePoint Foundation Incoming E-Mail Microsoft SharePoint Foundation Subscription Settings Service Microsoft SharePoint Foundation Timer *Microsoft SharePoint Foundation Tracing* *Microsoft SharePoint Foundation Usage* The Microsoft SharePoint Foundation web application Microsoft SharePoint Foundation Workflow Timer Service Microsoft SharePoint Insights The PowerPoint Conversion service The *Portal* service Project Server Application Service Request Management Search Administration Web Service The Search Host Controller service Search Query and Site Settings Service Secure Store Service *Security Token Service* SharePoint Server Search *SSP Job Control Service* User Profile Service Word Automation Services

While the indicated services are normally hidden, they can be viewed through SharePoint Management Shell with the `(Get-SPServer <server_name>).ServiceInstances` command.

Knowing which services are grouped together will help guide your selection of farm topologies, which we'll look at next.

Farm topologies

A farm topology is used to describe a type of SharePoint farm. There are three core SharePoint farm topologies:

- **Content farm**: A content farm hosts sites and service applications. It's the most common and broad type of farm. Content farms optionally consume services applications from other farms.
- **Services farm**: A services farm hosts service applications that are consumed by other farms.
- **Search farm:** This specialized service farm hosts **Search** service applications that are used by other farms.

Each farm type requires a different mix of server roles in order to function correctly. The following table describes which roles are required for which type of farm:

Server Role	Required for Content Farm?	Required for Services Farm?	Required for Search Farm?
Front-end	X	No	No
Application	X	X	No
Distributed Cache	X	X	No
Search	Only if hosting **Search**	Only if hosting **Search**	X

After selecting which type of farm you'll design, you can choose which of the server roles from the preceding table are appropriate for the farm. Then, once you have identified the type of farm you want to design and deploy, you'll need to address the high availability and disaster recovery concerns.

Designing for high availability

When designing any system for high availability, a number of questions/concerns are typically addressed, such as the following:

- What types of failures should a system be able to sustain?
- How many failures should a system be able to sustain?
- What steps (manual or automatic) need to be executed to ensure availability?
- What systems or processes can we put in place to avoid interruptions in the first place?

These types of questions speak to the concept of **dependability**. A dependable system is one that is available to service a request and is able to continue serving requests despite failures of the component architecture (such as a server or network device) or supporting services (such as electricity). Dependability has six core attributes:

- **Availability**: Measures the system's readiness to accept and respond to new requests for service
- **Reliability**: Measures how a system can continue to operate after an unexpected event
- **Safety**: Measures a system's level of risk to users and the environment
- **Confidentiality**: The ability to control or prevent unauthorized disclosure of information
- **Integrity**: Measures the presence or absence of an improper system alteration (such as data corruption)
- **Maintainability**: A qualitative measurement for how easily a system is kept current, repaired, or updated

When designing a system, these ideas or attributes of dependability can be used when building a Fault-Error-Failure chain to help identify potential errors and solve them before they are expressed during operation.

 The Fault-Error-Failure chain design principles are used in the development of most modern, highly available systems. The original work that introduces this, *Fundamental Concepts of Dependability*, is available at https://www.cs.rutgers.edu/~rmartin/teaching/spring03/cs553/readings/avizienis00.pdf.

From a practical standpoint, these questions of dependability can be broken up into four main categories:

- Fault forecasting
- Fault avoidance
- Fault removal
- Fault tolerance

Let's examine each of these with regard to designing a highly available SharePoint Server environment.

Fault forecasting

Fault forecasting is the prediction of likely or potential failures. With respect to SharePoint Server architectures, some of the following components come to mind:

- Server hardware, including components such as memory, chassis, power supplies, or mainboards
- Storage hardware, including components such as disk drives or other storage media, storage array software or firmware, or disk controllers
- Networking, including device (switch, router, firewall, proxy, and load-balancers) and cabling components, and inbound and outbound connectivity to the internet or other sites
- Power, including any power cables, switch boxes, outlets, power strips, uninterruptible power supplies, building or site power, and redundant power generation
- Software, such as application binaries or updates, **Secure Sockets Layer** (**SSL**) certificates, operating system binaries or updates, database servers, application services, and components

Each of those component categories represents one or more potential failures for an environment. In the forecasting stage, it's important to determine as many things as possible that can go wrong, as well as the likelihood and service impact of each.

Faults *will* happen in any environment, so devising strategies to identify potential faults and their impacts will help you design highly available systems.

Fault avoidance

Once potential faults in architecture have been identified, you can design around them. The premise of fault avoidance (or fault prevention) is to introduce elements that prevent faults. In the context of SharePoint Server architecture, this can mean several things, such as the following:

- Rigorous change control processes to understand modifications being made to the environment
- Development, test, or other sandbox-style environments where modifications are made and evaluated prior to production deployment

- Automated or scripted procedures to reduce the opportunity of human-caused failures
- Planning for redundancy and multiple failure modes

Fault avoidance is critical from both the design and operational perspectives to help ensure a high level of service and availability for a given service or application.

Fault removal

The goal of fault removal is to reduce the number and severity of service faults. Fault removal activities can be broadly divided into two categories:

- During the planning, design, or development of a system
- During the operation of a system

From a SharePoint Server perspective, removing faults during the development or planning of a system is the iterative process of identifying potential faults, such as disk drive or database failure (fault forecasting), designing a system to mitigate or prevent them (fault avoidance), and then performing testing that would trigger a particular failure mode.

For example, if you are planning for disk drive failure in a storage array, you would do the following:

1. Implement a storage subsystem with redundant features, such as disk mirroring.
2. Deploy an application or service utilizing the storage subsystem.
3. Introduce a failure, such as removing a disk drive, that would normally trigger a system failure.
4. Verify that the application or service continues to operate.

If the service or application fails to continue operating, you need to review the error logs and conditions, revise the deployment methodology or design, and then repeat the testing. Through this process, you can provide assurance to the business that the system will perform as designed.

Addressing the concept of fault removal during operation, using the previous example of disk drive failure, might look something like this:

1. The disk in the storage subsystem fails.
2. The disk subsystems continue operating in a degraded state.
3. The technician replaces the failed disk.
4. The system returns to a normal operational state.

In the preceding example, *Step 1* is the failure mode. *Step 2* indicates that the system's design has successfully resulted in continuing operations. In *Step 3*, the technician is performing fault removal by removing a failed device and replacing it with an operational one. In *Step 4*, the system has recovered and has returned to a normal operating state, free of faults.

In the previous failure scenario, the disk subsystem may have been designed to sustain the failure of a single disk drive. After the disk has failed in *Step 1*, the system is then at risk until the disk has been replaced in *Step 3*. The ability for a system to continue operation is compromised with each further fault, so it's important to minimize the amount of time between the steps.

Fault tolerance

Finally, the design goal of fault tolerance is to address how systems react *when* faults happen. As we've already stated, faults will happen. Fault-tolerant design plays a crucial role in allowing services to continue while faults are removed.

As a practitioner, you'll often be faced with choices and trade-offs to make on fault-tolerant designs, such as spending resources on redundant database hardware or additional servers in the SharePoint Server farm.

When designing highly available, fault-tolerant design for SharePoint, you'll likely need to incorporate the following components:

Fault Domains	Examples
Rack and power infrastructure	Server racks, power distribution units, power circuits, uninterruptible power supplies, fans, and cooling equipment
Physical server infrastructure and components	Servers, server chassis, server backplanes or midplanes, hard disk drives, controllers, network interface cards, and processors
Virtual server infrastructure and components	Virtual machine hosts
Network infrastructure and components	Rack-based switches, cabling, core switching, load balancers and traffic directors, and firewalls
Storage infrastructure and components	Storage networking components, disk arrays, disks, disk controllers, and **Redundant Array of Independent Disks (RAID)** settings.

Fault Domains	Examples
Application services and components	SharePoint application servers, Distributed Cache servers, User Profile Service, and the Search Service application
Database services and components	The SQL Server database failover clustering or AlwaysOn availability groups for content, configuration, and service application databases

In the fault forecasting step, you identified potential failures that could affect the SharePoint Server system and designed methods in the fault avoidance step to help mitigate or reduce the impact of the faults on the environment.

In addition to fault-tolerant designs, you also need to make preparations for how to recover from catastrophic failures (such as a natural disaster) that spans all components in either a single fault domain or multiple fault domains.

In the next section, we'll look at using highly available designs to mitigate the impact of failures of various service databases.

Supported SharePoint high-availability designs

A SharePoint farm has many moving pieces. A successful highly available design requires understanding how the various components can be made resilient. The following table lists the database design considerations:

Service Database	Supports Database Mirroring for High Availability	Supports Database Mirroring or Log Shipping for Disaster Recovery	Supports SQL AlwaysOn Availability Group for Availability	Supports SQL AlwaysOn Availability Group for Disaster Recovery
Configuration database	X		X	
Central Administration database	X		X	
Content database(s)	X	X	X	X
App Management database	X	X	X	X
Business Connectivity Service database	X	X	X	X
Managed Metadata Service database	X	X	X	X

Service Database	Supports Database Mirroring for High Availability	Supports Database Mirroring or Log Shipping for Disaster Recovery	Supports SQL AlwaysOn Availability Group for Availability	Supports SQL AlwaysOn Availability Group for Disaster Recovery
PerformancePoint Services database	X	X	X	X
Power Pivot Service database	X	X	X	X
Project Server database	X	X	X	X
SharePoint Search Service – administration database	X		X	
SharePoint Search Service – analytics reporting database	X	X	X	
SharePoint Search Service – crawl database	X		X	
SharePoint Search Service – link database	X		X	
Secure Store database	X	X	X	X
SharePoint Translation Services database	X	X	X	X
State Service database	X			
Subscription Settings database	X	X	X	X
Usage and Health Collection database	X	X	X	
User Profile Service – profile database	X	X	X	X
User Profile Service – synchronization database	X	X	X	X
User Profile Service – social tagging database	X	X	X	X
Word Automation Services database	X	X	X	X

For more information on the specific SQL or SharePoint versions necessary to support certain high-availability designs, go to `https://docs.microsoft.com/en-us/sharepoint/administration/supported-high-availability-and-disaster-recovery-options-for-sharepoint-databas`.

One of the common threads you'll see in the databases' availability design is the support for SQL Server AlwaysOn availability groups. Microsoft recommends AlwaysOn availability groups for all databases in a SharePoint Server environment from the perspective of same-farm high availability.

Service Applications support high availability behind load-balancers. After using the SharePoint product configuration wizard to configure a role for your server, add a configuration object (such as a virtual IP) to your load balancer that includes all of the servers hosting an application or service.

While a fault-tolerant and resilient design is important from a design and day-to-day operational perspective, you also need a plan for business continuity concerns in the event of a significant problem. That is where disaster-recovery planning is helpful.

Planning for disaster recovery

Disaster recovery is the set of measures you undertake when your deployment has undergone a significant failure that exceeds the capabilities of your fault tolerance. Some example scenarios that might require disaster recovery efforts include the following:

- **Storage failures**: For example, if your storage environment has two redundant disk controllers and both of them fail before you can return the system to full capacity, or more than one disk fails simultaneously in a RAID-5 disk volume.
- **Virtual machine host failures**: If your environment comprises virtual machines and the underlying virtual machine hypervisors fail in a way that prohibits the virtual machines supporting your environment from powering up.

- **Software updates**: This could apply to operating system updates, application platform updates, driver updates, or other application updates that render the system unusable.
- **Database failure**: Since the majority of SharePoint Server's services rely on storing and retrieving information from databases, a catastrophic database failure could prohibit components in the farm from working correctly.
- **Primary data center site compromise**: Any event that impacts your primary data center, such as extended power outage, a flood or another natural disaster, network connectivity service interruption, or military action.

Your organization may require you to be prepared to resume activities in the event of any of these scenarios (or others that may apply to your environment). The ability to recover or restore operations is gauged by three measurements:

- **Recovery Point Objective (RPO)**: The RPO can be expressed in several ways, such as "the last available backup from which to initiate a restore" or "the acceptable amount of data loss."
- **Recovery Level Objective (RLO)**: A sub-function of the RPO, the RLO defines the granularities that you need to be able to recover (such as a data center, rack, host, farm, server, application, database, site, document library, folder, or file).
- **Recovery Time Objective (RTO)**: The amount of time it takes to get a system operational with the data parameters of the RPO. This can also be referred to as how long the outage can last or "how long we're down."

 When starting to develop an RPO, many organizations state that "no data loss is acceptable." While no-loss solutions are possible, the more data a system contains and a higher frequency of activity could significantly impact the overall cost of a solution. Frequently, a "no data loss" policy is not cost-justifiable. The business needs to determine how valuable an outage is (quantified by the business, legal, and financial risk from an extended outage) before work can begin on recommending technical solutions.

A business recovery objective or requirement might be expressed as follows:

Must be able to recover a SharePoint farm at the document library level (RLO) in less than 2 hours (RTO) with no more than 2 hours of potential data loss (RPO).

As you put together a disaster recovery plan for SharePoint Server, it's important to start with the organization's goals (such as the number of hours of downtime or how much potential data loss is acceptable), and then recommend strategies, processes, and products based on that business requirement.

Outage costs

Outages fall into three categories, generally as follows:

- Planned loss of application or service (such as a service upgrade or scheduled maintenance)
- Unplanned loss of application or service
- Loss of data

Loss of an application or service may prohibit your organization from generating revenue or performing required activities for the business to operate, which may have a financial impact, depending on the application or service that is inoperable. An application or service can also incur a *partial* loss (such as running in a degraded fashion), which may render the system usable for some activities and not for others.

Planned outages are typically communicated to business users or customers and are scheduled to happen during low periods of activity. Unplanned outages, conversely, happen without notice due to some type of system failure.

Loss of data, depending on the type of data affected, could have a significant financial impact on an organization.

Depending on the type of application or data hosted by a SharePoint Server environment and the type of outage incurred, you may need to evaluate one or more disaster recovery options.

Disaster recovery options, costs, and considerations

Disaster recovery options (and their costs) can be quite varied, from a simple backup and restore to full standby data center solutions. Here are some example of disaster recovery options:

Type	Components	Notes	Relative Deployment and Maintenance Cost	Recovery Time
Tape or disk-to-disk backup solution	Tape or disk-to-disk backup hardware, software	This simplest form of disaster recovery covers only the applications and data. It is typically the cheapest option to deploy and maintain, but it depends on the organization being able to provide infrastructure, should the need to recover data arise.	Lowest	Longest
Cold standby infrastructure	Dedicated servers ready to be configured in the event of a disaster	This solution builds on having backups by providing dedicated hardware. This hardware is not configured or maintained but is waiting for a disaster so that it can be configured to meet the exact recovery requirements. Cold standby infrastructure is typically infrastructure that can be available within hours or days.	Low	Long

Type	Components	Notes	Relative Deployment and Maintenance Cost	Recovery Time
Warm standby infrastructure	Dedicated servers that are regularly maintained and available	A warm standby infrastructure disaster recovery scenario leverages dedicated equipment that is kept up to date on a schedule using regular restores or synchronizations of data. Warm standby infrastructure can typically be used to make a solution available within minutes to hours.	Medium	Medium
Hot standby infrastructure	Dedicated servers that are regularly maintained and kept up to date, ready for failover	Hot standby infrastructure, like warm standby infrastructure, is dedicated equipment that is kept up to date. Unlike warm standby infrastructure, however, hot standby infrastructure is ready to take over within seconds to minutes. Hot standby infrastructure plans frequently rely on load balancing and data replication technologies.	Expensive	Shortest

Type	Components	Notes	Relative Deployment and Maintenance Cost	Recovery Time
Cold standby data center	Dedicated data center space with equipment ready to be provisioned	A cold standby data center strategy relies on having available equipment and backups at a secondary location. This is a somewhat expensive solution to maintain (a data center space and networking and server equipment is required, as well as ensuring backups are available) and has both high-recovery time and point objectives. It will likely take days or weeks to get a cold standby data center operational.	Somewhat expensive	Long
Warm standby data center	Dedicated data center space with pre-configured equipment, ready to accept failover or restores	Similar to a warm standby infrastructure solution, a warm standby data center disaster recovery solution means you have equipment mostly up to date at a remote location. The most recent data can be applied to this environment, typically within minutes or hours.	More expensive	Medium

Type	Components	Notes	Relative Deployment and Maintenance Cost	Recovery Time
Hot standby data center	Dedicated servers that are regularly maintained and kept up to date, ready for failover in a separate data center space	Building on the concepts of hot standby infrastructure, a hot standby data center recovery strategy is the most resilient (and expensive) solution to maintain as it requires both investment (data center space, dedicated equipment, software, networking, and communications) and sound process execution. Hot standby data centers can be ready within seconds to minutes and can have the lowest recovery time and recovery point objectives for overcoming full primary site disaster.	Most expensive	Shortest

As with designing a fault-tolerance strategy, you'll also want to design a disaster recovery strategy that takes failure domains into account. These failure domains might include the following:

- Application, workload, database, or service
- Infrastructure or platform
- Farm
- Data center

Finally, no disaster recovery plan is complete without documentation that allows the technicians or support staff to return services to their full operational status. These operational recovery plans (sometimes referred to as runbooks or playbooks) should include things such as the following:

- Step-by-step printed instructions used to recover services from each failure or disaster mode, such as operating system installation and configuration, configuration, IP address schemes, or database names
- Tested scripts for building, deploying, and testing the configuration
- Operational procedures for restoring data
- Correct versions of software installation media and any applicable licensing information (such as key files, licenses, or other activation/registration information necessary to bring the service online)
- Emergency contact information for building access, infrastructure personnel, and application or business owners

Evaluating the business objectives (recovery time objective and recovery point objective) in conjunction with the budget will help you arrive at an appropriate disaster recovery strategy for your organization.

Azure Site Recovery is a Microsoft Azure-based disaster recovery service that can be leveraged in lieu of building and maintaining a physical disaster recovery site. It can be used as a disaster recovery solution for physical or virtual machines. For more information on configuring Azure Site Recovery for SharePoint, go to `https://docs.microsoft.com/en-us/azure/site-recovery/site-recovery-sharepoint`.

Next, we'll look at backup and restore as part of the SharePoint Server planning process.

Planning and configuring backup and restore options

Backups are an important part of designing and maintaining a SharePoint Server environment. The ability to backup and recover data is integral to many business processes, such as the following instances:

- Recovering or restoring accidentally or maliciously deleted content
- Platform or service upgrades

- Migrating between infrastructure solutions
- Recovering from failed upgrades
- Recovering from unexpected system failures or disasters

Natively, SharePoint Server supports the following backup designs:

- Farm
- Granular
- Configuration only

Let's examine these backup designs in more detail.

Farm

In a farm backup scenario, you'll be targeting backing up major components of the farm. The farm backup architecture is initiated from Central Administration and launches a SQL Server backup of content and service application databases. It also saves configuration content to files and backs up the **Search** index data.

A SharePoint farm backup supports both **Full** (all data) and **Differential** (data that's changed since the last full backup).

Within a farm backup, you can select the following nodes:

- **Farm**: The farm is the highest-level object. You can choose from the following selections:
    - Content and configuration data (default), which backs up the entire server farm (including settings from the configuration database)
    - Configuration-only, which only backs up the settings stored in the configuration database
- **Web application**: The **Web application** node represents the pieces necessary for an entire web application, including the following:
    - App pool name and account information
    - Authentication settings configuration
    - General web application settings, such as managed paths

- **Internet Information Services** (**IIS**) site binding configuration, such as the protocol, port, and host header information
- Changes to the `Web.config` file that were made through object model-based scripting or applications, as well as changes made through Central Administration (changes made manually are not backed up)
- Sandboxed solutions (sandboxed solutions are stored in a web application's content databases)

- **Services and service applications (not shared)**: Service and service application backups contain the settings for that service or service application, as well as any databases associated with it. If backing up a service application, the related proxy is *not* included. To fully protect the service application and its related application proxy, perform either a full farm backup or two consecutive backups with the service application in the first and the application proxy in the second.

- **Proxies for service applications that are not shared**: This is the application proxy only.

- **Shared Services**: Shared services require both a service application and an application proxy to run. Choosing the **Shared Services** node will back up all of the service applications and the corresponding service application proxies.

In addition, there are things that even a full backup doesn't cover (mostly related to the operating system and IIS configurations):

- Application pool account passwords
- HTTP compression settings
- Time-out settings
- Custom **Internet Server Application Programming Interface** (**ISAPI**) filters (configured through IIS)
- Computer domain membership, group membership, or organizational unit placement
- **Internet Protocol security** (**IPsec**) settings
- **Network Load Balancing** (**NLB**) settings
- SSL certificates or certificate trust stores
- Dedicated IP address configurations and settings
- Windows Firewall settings

For these settings, you will likely want to perform operating system backups or exports. Furthermore, there are a few additional caveats for farm-based backups that you should be aware of when planning a backup and restore strategy:

- The native SharePoint Server backup tools do not have a native scheduling interface. To perform scheduled backups, you will need to create a backup script of your own and schedule it using Windows Task Scheduler.
- SharePoint Server backup does *not* protect changes to `Web.config` made outside Central Administration or object model-based scripting solutions.
- SharePoint Server backup does *not* protect individual site customizations that are not part of a trusted or sandboxed solution.
- SharePoint Server backup does *not* protect trust certificates that have deployed for cross-farm solutions or service applications.
- Databases configured to use SQL FILESTREAM **Remote Blob Storage** (**RBS**) can only be backed up and restored to SQL database servers with the RBS provider installed.
- If your farm or applications use forms-based authentication, those settings must be manually configured after you perform a restore by registering the membership and role providers and redeploying the providers.

Any flat files that you backup through PowerShell or another export process can then be backed up using software such as System Center Data Protection Manager.

Granular

When backing up certain components (such as sites, workflows, customizations, web applications, or other individual components), you perform a granular backup. As you saw in the previous section, there are a number of tools you can use to perform various types of backups.

A granular backup, when using native tools, relies on Transact-SQL statements to export content. With the granular backup system, you can export or backup site collections and lists.

 Workflows are not included when performing exports of sites or lists.

You can perform granular backups using native tools, including PowerShell, Central Administration, and SQL Management Studio. If you are using a version of SQL that supports database snapshots (such as Enterprise Edition), backups can trigger a database snapshot and allow users to continue accessing the site as normal. When backup utilizes database snapshots, you gain the added advantage of a totally consistent backup at a particular point in time. The downside is that it is a more I/O-intensive operation to perform a database snapshot, so users may experience a brief slowdown when the database snapshot is initiated.

 As previously mentioned, the native SharePoint backup solution administration is performed from the farm's Central Administration page. Many organizations, however, use third-party backup solutions that utilize an application-aware backup agent for SharePoint services, applications, and databases.

Whichever method you choose for performing a backup (whether you are using full farm backups or more granular backups), you'll have a route to restore your environment or its components should unexpected circumstances arise.

Configuration only

With a configuration-only backup, you are only backing up the server or farm configuration without any data. This type of backup might be useful for replicating farm configurations in conjunction with disaster recovery preparation or building test and development environments. Depending on the mechanism used, you can back up either the entire farm configuration or parts of the farm or server configuration using either the Central Administration interface or by using PowerShell.

Central Administration

By performing a configuration-only backup with SharePoint Server Central Administration, you can back up the configuration of the farm. The Central Administration configuration backup method can *only* be used to back up the full configuration of the local farm. It *cannot* be used to back up a remote farm or a disconnected configuration database.

To perform a farm backup using Central Administration, follow these steps:

1. Verify that the user account performing the backup is a farm administrator.
2. Navigate to the **Central Administration** page. Under **Backup and Restore**, click **Perform a backup**.
3. On the **Perform a Backup — Step 1 of 2: Select Component to Back Up page**, select the farm from the component list browser, then click **Next**.
4. On the **Start Backup — Step 2 of 2: Select Backup Options** page, select **Full** in the **Backup Type** section.
5. In the **Backup Only Configuration Settings** section, select **Backup only configuration settings**.
6. In the **Backup File Location** section, enter the **Universal Naming Convention** (**UNC**) path of the destination backup folder, then select **Start Backup**.

The UNC path will contain a backup that can further be moved to offline media and preserved.

PowerShell

When using the `Backup-SPConfigurationDatabase` PowerShell cmdlet to perform a configuration-only backup, you can overcome the limitations of performing a backup with Central Administration. PowerShell-based backups allow you to backup configurations from either local or remote farms, as well as disconnected configuration databases.

Performing the backup from PowerShell, depending on the target configuration database, may require additional permissions. Generally speaking, the account that is used to perform the backup must have the following permissions or memberships:

- A `db_owner` fixed database role on all of the databases being backed up or updated
- A `securityadmin` fixed server role on the SQL server instance where the databases are hosted
- Membership to the local **Administrators** group on the server where `Backup-SPConfigurationDatabase` is being executed
- Membership to the `WSS_ADMIN_WPG` local group on the server(s) where SharePoint products are installed
- A `SharePoint_Shell_Access` role for the databases that are being acted upon
- Optionally, the `db_backupoperator` fixed database role for the databases being backed up or updated

You can use the `Add-SPShellAdmin` cmdlet to grant the account used for PowerShell administration access to the `SharePoint_Shell_Access` role to specific databases. It is recommended that you specify which databases to use with the `-Database` parameter as that will add the named account to the farm configuration database, the Central Administration content database, and the database specified in the parameter, as well as to the `WSS_ADMIN_WPG` group on all SharePoint web servers.

You can use a function or PowerShell script, such as the one in the following code block, to accomplish the task of adding an additional administrator to the SharePoint farm. Copy and paste the following script into SharePoint Management Shell, then run `New-FarmAdmin -Identity <username> -IncludeAllContentDatabases`:

```
Function New-FarmAdmin
([string]$Identity,[switch]$IncludeAllContentDatabases)
{
    $CentralAdminWebApp = Get-SPWebApplication
-IncludeCentralAdministration | `
    ? {$_.DisplayName -like "SharePoint Central Administration*"}
    New-SPUser -UserAlias $Identity -Web $CentralAdminWebApp.URL -Group
"Farm Administrators"
    $CentralAdminContentDB = Get-SPContentDatabase -WebApplication
$CentralAdminWebApp
    Add-SPShellAdmin -Database $CentralAdminContentDB -Username $Identity
    If ($IncludeAllContentDatabases)
    {
        $ContentDatabases = Get-SPContentDatabase
        Foreach ($Database in $ContentDatabases)
        {
            Add-SPShellAdmin -Database $Database -Username $Identity
        }
    }
}
```

 For more information on the Add-SPShellAdmin cmdlet, see `https://docs.microsoft.com/en-us/powershell/module/sharepoint-server/Add-SPShellAdmin?view=sharepoint-ps`.

You must also have write access to the network share path that will be used as the destination for the backup. The `Backup-SPConfigurationDatabase` cmdlet is used to initiate a backup.

To perform a configuration backup using PowerShell, perform the following steps:

1. Launch **SharePoint Management Shell**.
2. Run `Backup-SPConfigurationDatabase -Directory <BackupFolder> -DatabaseServer <DatabaseServerName> -DatabaseName <DatabaseName>`.

Be sure to specify the `-Directory` parameter value destination as `\\server\share`.

Both planning for disaster recovery scenarios as well as day-to-day backup and restore scenarios are crucial to ensuring your data is protected from destruction or environment failure. In the next section, we'll discuss how **Information Rights Management** (**IRM**) can be used to protect your information in another way—from unauthorized use, tampering, or leakage.

Planning for IRM

IRM is the set of policies and protections used to govern the actions that users can take on documents stored in SharePoint. IRM relies on some form of Rights Management Services. SharePoint Server, depending on the version, supports the use of Active Directory Rights Management Services (all on-premises versions) or the corresponding cloud version, Azure Rights Management (through the use of the Rights Management Services connector for SharePoint Server 2010, SharePoint Server 2013, and SharePoint Server 2016) to protect assets. Additionally, Azure Information Protection is a rights management-based protection that can be applied at the individual file level (whether a file is in a SharePoint library or not), making it an ideal method for protecting data across an enterprise.

Determining which solution is best for your organization depends on both your current infrastructure and knowing what the technology roadmap for your organization entails. Most organizations will leverage more cloud services over the course of time, so it is best to understand what direction your particular organization will take.

While it is important to understand that both the Active Directory Rights Management and Azure Rights Management platforms are supported by SharePoint Server 2016, this book will focus on utilizing Azure Rights Management and Azure Information Protection. Both Azure Rights Management and Azure Information Protection are available cross-premises.

All of the rights management protection schemes allow document owners (or document library owners) to protect supported documents—typically, this includes Microsoft Office formats as well as the **XML Paper Specification** (**XPS**). The information protection technology that is deployed determines what file formats and types can be protected. For example, native SharePoint Server 2016 or 2019 IRM only supports Office file formats and XPS, while Azure Rights Management supports many additional common document formats. For more information on the exact file formats supported, see `https://docs.` `microsoft.com/en-us/azure/information-protection/rms-client/client-admin-guide-` `file-types`.

Depending on the product that is implemented (Active Directory Rights Management, Azure Rights Management, or Azure Information Protection) and the location (document library, document, or email), some or all of the features, permissions, or usage rights may be able to be applied:

- **Edit**: Allows a user to modify the content stored in a document's associated application. It does not natively grant the ability to save the document.
- **Save**: Allows the user to save the document to the current location. Depending on the application, users may also be able to save the document to a new location.
- **Comment**: Enables the option to add comments or annotations.
- **Save As** or **Export**: Allows the user to save the content to a different filename or export content. This also supports exporting content to different applications (such as **Send to OneNote**).
- **Forward**: Enables the user to forward an email to additional users and modify the **To** or **Cc** lines. *Note that this only applies to an actual email message and has no bearing on the rights present in any attached document.*
- **Full Control**: Enables all rights to a document, including the ability to add or remove protections and restrictions.
- **Print**: Enables the option to print content.
- **Reply**: Enables the ability for users to reply to the sender of a rights-protected email.
- **Reply All**: Enables the ability for users to reply to all **To** or **Cc** recipients in a rights-protected email.
- **View**, **Open or Read**: Allows the user to open a document or email and see the content. This does not allow users to change the contents of a document (for example, sorting or filtering a column in Excel). Clicking on the content in a protected document frequently requires some form of the **Edit** permission.
- **Copy**: Allows the ability to copy the content or perform screen captures.

- **View Rights**: Enables a user to view the rights assigned to a document.
- **Change Rights**: Enables a user to change the rights policy applied to a document, including the ability to remove all protection from a document.
- **Allow Macros**: Enables a user to run a macro or enable other programmatic access in a protected document.

With that being said, when planning for IRM, you'll need to understand the SharePoint environment, what software clients or applications will be used, where the rights need to be applied, and which supported technologies are available.

Planning localization and language packs

Part of any SharePoint Server deployment instance is understanding the intended target audience. SharePoint Server supports some level of localization. Localization is the concept of SharePoint displaying content in a different language than it may have been created in. Multilingual features enable users to see sites in their preferred languages.

SharePoint Server supports two core types of localization:

- Multilingual user interface
- Variations

In the next two sections, we'll explore the features and use cases for both of them.

Multiple Language User Interface

With the **Multiple Language User Interface** (also referred to as the **Multilingual User Interface** or **MUI**), you can configure SharePoint Server to display interface elements in a user's preferred language. MUI configurations affect the elements that are used to manipulate and interact with SharePoint, such as navigation items. When using the MUI, you can display the following user interface elements in different languages:

- Default columns
- Navigation bar links
- SharePoint default menus and actions
- Custom columns for a list or site
- The site title and description
- The Managed Metadata services

When planning to use MUI features, you'll need to understand the languages that your users commonly use. MUI features require the installation of language packs for the languages that you want to use.

MUI features are configured at the site level. To change them, follow these steps:

1. As a site collection owner or administrator, navigate to the site where you wish to enable multiple language support.
2. Click **Settings** | **Site Settings**. Depending on whether the site is on the modern or classic version or has been customized, you may need to select **Site Information** | **View all site settings** on the **Site settings** fly-out menu.
3. On the settings page, under the **Site Administration** section, click **Language Settings**.
4. Select the checkboxes for the languages you want to make available.
5. Click **Yes** under **Overwrite Translations** if you want to overwrite translations in the site interface when changes are made to the default site interface. If other administrators make changes in alternative language versions of the page (such as updating the translated text for a navigation item), selecting **Yes** on this setting will cause the changes made in the default interface to overwrite the customized changes in an alternative language interface.
6. Click **OK**.

It's important to note that while *navigational items* or *list columns* will be translated, the *content* will not.

Variations

While the MUI is used to translate navigational and interface elements, **variations** can be used to actually translate content. This can be useful if you have content that needs to be replicated in several languages. Variations require sites to be built with the **Publishing Site** templates, which includes workflows and timer jobs to update pages.

Variations rely on the configuration of several elements:

- **Variation root site**: The root site provides the URL for the variation sites. It hosts the landing page that will redirect users to the correct site.
- **Variation labels**: An identifier for a variation. Each unique variation needs to have its own variation label defined.
- **Variation sites**: Sites that are created and managed based on the variation labels.

- **Source variation sites**: Denotes sites where content is authored, curated, and published. A source variation site's content is synchronized with the target variation sites. A site collection can only host one source variation site.
- **Target variation sites**: Denotes sites that receive synchronized copies of content from source variation sites. Target variation sites can also host new, unique content (though that content is not synchronized to other sites).
- **Variations hierarchy**: A complete set of sites with all variation labels.
- **Variation lists**: Lists for which you identify target variation labels to receive list items.
- **Variation pages**: Publishing pages stored in the **Pages** library of the source and target variation sites.

When planning a variations-based deployment, there are several important things to consider:

- **Content approval**: As previously noted, the use of variations requires SharePoint Server **Publishing Site** templates. In order for new content to be synchronized and made available on target variation sites, it must be *approved* by the source variation site administrator(s) or member(s) with approval permissions. As content is only visible once it has been approved, part of the planning process needs to focus on who is responsible for both creating and approving content to be pushed to sites that will have variations. Variations and content approval require major and minor versioning enabled in the source and target variation sites.
- **Content deployment**: Content deployment jobs are used to copy content from one site collection to another. If you plan on using content deployment jobs to sync content into a source variations site, you need to ensure that the content deployment job schedule doesn't overlap with the **Variations Create Hierarchies Job Definition** job. If they execute at the same time, you risk inconsistent site data being synchronized from the source to target variation sites.
- **Cross-site publishing**: Cross-site publishing is a feature that allows content from one site to be shown in another site's search results. The source (or authoring) site collections are used to author and contain content and then a publishing site is used to control the design and show the content. The authoring sites contain **catalogs** (content that is tagged with various metadata). The publishing site collection uses **Search** web parts to display cataloged content based on the filters the site author selects. There are three core scenarios for cross-site publishing in variations or multilingual sites, each of which has their own architectural and planning guidance (see `https://docs.microsoft.com/en-us/sharepoint/administration/plan-variations-for-multilingual-cross-site-publishing-site` for more detailed deployment and configuration information):

- Maintaining multilingual content outside of SharePoint
- Only publishing multilingual content, or publishing a mix of catalog- and non-catalog-based content, without variations on the non-catalog content
- Publishing a mix of catalog and non-catalog content with variations on all content

- **Site navigation**: Like cross-site publishing, site navigation in variations-based sites has additional planning requirements. Site navigation is automatically generated and displayed in the **Global Navigation** and **Current Navigation** menus of a web page. Depending on your layout, this may or may not be a desirable feature.
- **Web parts**: Web parts are used to display content on SharePoint pages. Web parts are synced by default as part of the variations configuration. Depending on the type of content displayed in a web part, it may be desirable to disable web part updates to target variations sites.

It's important to note that the engine Microsoft previously used for machine translation (`https://docs.microsoft.com/en-us/sharepoint/dev/general-development/machine-translation-services-in-sharepoint`) has been deprecated but will continue to be supported, per the release notes of SharePoint Server 2019 (`https://docs.microsoft.com/en-us/sharepoint/what-s-new/what-s-deprecated-or-removed-from-sharepoint-server-2019`). As indicated at `https://support.office.com/en-us/article/create-a-multi-language-website-da0b5614-8cf5-4905-a44c-90c2b3f8fbb6`, Microsoft has recommended the use of Azure Cognitive Services (formerly the Bing Translator API), but does not provide many resources on configuring or using alternative services. There are a few third-party products (both free and paid) that will allow you to perform translation services if necessary.

Regardless of the translation mechanisms that are used, it's important to determine what languages your users need and what mechanisms you will use (manually creating and updating content, MUI items, and custom or third-party translation services).

Localization settings will help you tailor your content to the language needs of your individual users. In the next section, we'll talk about how and where to store the content and applications themselves—in SharePoint Server content farms.

Planning and configuring content farms

For many organizations, a single content farm is the most practical and simple choice to deploy. Utilizing some of the design concepts discussed earlier in this chapter (MinRole farm topologies and high availability), you can choose from some of the recommended design patterns, depending on your organization's budget and performance needs:

Small High Availability MinRole-based farm (4 servers)	Four servers with two shared roles: • Two Front-end shared with Distributed Cache • Two Application shared with Search	Front-end & Distributed Cache	Application & Search		
		Front-end & Distributed Cache	Application & Search		
Medium High Availability MinRole-based farm, Search optimized (6 servers)	Six servers: • Two Front-end shared with Distributed Cache • Two Application servers • Two Search servers	Front-end & Distributed Cache	Application	Search	
		Front-end & Distributed Cache	Application	Search	
Medium High Availability MinRole-based farm, User optimized (6 servers)	Six servers: • Two Distributed Cache servers • Two Front-end servers • Two Application shared with Search servers	Distributed Cache	Front-end	Application & Search	
		Distributed Cache	Front-end	Application & Search	
Large High Availability MinRole-based farm (8 servers)	Eight servers: • Two Distributed Cache servers • Two Front-end servers • Two Application servers • Two Search servers	Distributed Cache	Front-end	Application	Search
		Distributed Cache	Front-end	Application	Search

Utilizing one of the built-in MinRole server roles (selected in the SharePoint Products Configuration wizard), you can choose where you want to invest resources, ranging from four highly available servers, each running two shared roles, and up to eight (or more) servers configured with dedicated roles. It is important to understand the MinRole deployment topologies and the components required for high availability from a design perspective to ensure you're in a supported design pattern.

In the next section, we'll look at extending the on-premises topology into Office 365.

Planning integration with Office 365 workloads

While SharePoint Server has a very large feature set, you may need to connect an on-premises SharePoint environment to Office 365—whether that's extending your environment out to Office 365 or accessing on-premises content from Office 365. The following features can be integrated between SharePoint Server and Office 365 components:

- **OneDrive for Business**: Hybrid OneDrive for Business configures users to be redirected to OneDrive for Business in Office 365 when they access OneDrive in SharePoint Server.
- **Sites**: Hybrid sites allow you to integrate navigation between SharePoint Server and Office 365.
- **Search**: Hybrid search scenarios allow you to configure a search environment that allows users to locate content in Office 365 or local SharePoint farms from either search environment.
- **Business Connectivity Services**: Allows access to on-premises data using Business Connectivity Services.
- **Power Apps, Power BI, and Power Automate (formerly Microsoft Flow)**: Using a data gateway, you can connect the newest Office 365 Power Platform applications (Power Apps, Power Automate, and Power BI) to on-premises environments.

All of these features require running the SharePoint Hybrid Picker application to configure hybrid features, with the exception of configuring the data gateway for Power Platform integrations.

Choosing which types of integrations you need to deploy will depend largely on the business and use cases for your farm. For example, you may need to deploy a solution that requires SharePoint Server features and the ability to synthesize datasets that are in other on-premises systems, but also want users to be able to take advantage of Office 365's OneDrive for Business. Alternatively, you may have data in legacy on-premises farms that you're not planning to migrate to Office 365, but you want users to be able to find and access it from a single search view in Office 365.

An understanding of the use cases and any limitations for existing infrastructure largely determines what Office 365 workload integrations you will need to use.

Planning and configuring the OneDrive sync client

The OneDrive sync client is used to synchronize data, typically between a user's personal computer and their server-based OneDrive (also known as My Sites) document library.

The newest version of the OneDrive sync client works with both SharePoint Server and SharePoint Online, so depending on which one your organization uses (or plans to use), you may want to specify some additional configuration settings.

Planning Group Policy configuration settings

When planning your OneDrive sync client, there are a number of configuration items to address. Many of these configurations can be managed with Group Policy objects for the OneDrive client:

- **Specify the SharePoint Server URL and organization name**: When OneDrive for Business is configured against Office 365, a staging folder for the OneDrive client is created using the Office 365 tenant's name as a prefix. For example, if the tenant ID name is `contoso.onmicrosoft.com` and the organization is Contoso, Ltd., the OneDrive for Business folder would (by default) be named `C:\Users\<username>\OneDrive - Contoso, Ltd.`. On-premises works a little bit differently. SharePoint Server uses the prefix of the SharePoint URL to create the folder name. For example, if your SharePoint environment's name is `spfarm.contoso.com`, then the organization name that the OneDrive client will use is `spfarm` (and the resulting folder name would be `C:\Users\<username>\OneDrive - Spfarm`). Depending on your plans for Office 365 or OneDrive for Business in the future, you may wish to make the organization name reflect the value in an Office 365 tenant, if you have one. This setting is located under **Computer Configuration | Policies | Administrative Templates | OneDrive**:

Specify SharePoint Server URL and organization name   — □ ✕

Specify SharePoint Server URL and organization name   [Previous Setting]   [Next Setting]

◉ Not Configured   Comment:

○ Enabled

○ Disabled

          Supported on:

Options: Help:

Provide a URL to the SharePoint Server that hosts the user's OneDrive for Business, and the organization name.

SharePoint Server 2019 URL:

Organization name:

This setting lets you enable users to use the OneDrive sync app (OneDrive.exe) to sync files in SharePoint Server 2019. The URL defines the location of the SharePoint Server and enables the sync app to authenticate and set up sync. The organization name lets you specify the OneDrive and SharePoint folder names that will be created in File Explorer. The organization name is optional. If you don't provide it, the sync app will use the first segment of the URL as the name. For example, office.sharepoint.com would become "Office," and the OneDrive folder name would be "OneDrive - Office."

If you enable this setting and provide the SharePoint Server URL, users will be able to sync files in SharePoint Server 2019.

If you disable or do not configure this setting, or do not provide the SharePoint Server URL, users will not be able to sync files in SharePoint Server 2019.

[OK]   [Cancel]   [Apply]

- **Specify the OneDrive location in a hybrid environment**: If your organization has deployed both Office 365 OneDrive for Business and SharePoint Server, you may want or need to decide which personal storage service users will access. You can use this policy setting to determine whether users will authenticate to Office 365 (SharePoint Online) first or to a local SharePoint system. This policy object is located under **Computer Configuration** | **Policies** | **Administrative Templates** | **OneDrive**:

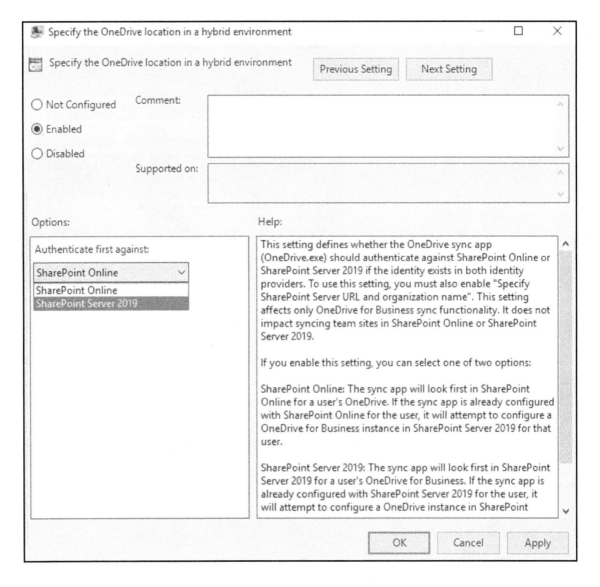

- **Use OneDrive Files On-Demand**: **Files On-Demand** is a new feature that allows you to save local disk space by storing a shortcut in the OneDrive sync folder to the content stored in SharePoint. Depending on your deployment constraints (such as non-persistent virtual desktops, devices with small local disk storage, or organizational policies that limit data storage on local devices), it may be preferable to use this feature to prevent the automatic synchronization of data when the OneDrive client starts. It requires Windows 10 Fall Creators Update (version 1709 or later), OneDrive sync app build 17.3.7064.1005 or later, and SharePoint Online or SharePoint Server 2019. The **Files On-Demand** feature is located under **Computer Configuration** | **Policies** | **Administrative Templates** | **OneDrive**.

In addition to these popular settings, there are several other OneDrive settings that may be useful when planning an OneDrive deployment. For a complete list of One Drive Group Policy settings, see `https://docs.microsoft.com/en-us/onedrive/use-group-policy`.

Deploying OneDrive administrative templates

If you have not yet deployed OneDrive template files previously in your organization, you can follow these steps to make the Group Policy settings available:

1. Download and install the most current OneDrive sync app for Windows. To obtain the latest version, go to `https://go.microsoft.com/fwlink/?linkid=823060`.

2. Once it has been installed, navigate
   to `%localappdata%\Microsoft\OneDrive\BuildNumber\adm`, where the
   build number is the latest version of the OneDrive app (which you can find on
   the **Settings** | **About** page of the OneDrive app):

3. Copy the `.adml` and `.admx` files.
4. Navigate to the SYSVOL policies container for your domain (typically
   `\\domain\sysvol\domain\Policies`) and expend the `PolicyDefinitions`
   folder. If the `PolicyDefinitions` folder does not exist, create it. In the
   `PolicyDefinitions` folder, create another folder for your language (for
   example, `en-us`).

5. Paste the `OneDrive.admx` file in your domain's `PolicyDefinitions` folder and
   paste the `OneDrive.adml` file in the specific language folder (for example,
   `PolicyDefinitions\en-us`).

Once the templates have been deployed, you can create a new Group Policy configuration for OneDrive and update the necessary policy objects.

In the next section, we'll discuss some of the steps you can take to get the best performance possible from your SharePoint environment.

Planning and configuring a high-performance farm

Virtualization is an enterprise for the architecture of most workloads and services in modern data centers. Virtualization brings a lot of benefits to computing, such as the improved capability to both horizontal and vertical scaling, to be able to add layers of host redundancy to support highly available solutions.

However, virtualization also introduces potential pitfalls, such as overcommitting hardware resources.

SharePoint Server is fully supported on virtualized platforms, as well as on traditional bare-metal platforms. Virtualization adds a layer of abstraction to the entire process, potentially making it more difficult to understand all of the components affecting your environment's performance.

The areas that we'll look at include the following:

- Core requirements
- Server infrastructure
- Processors
- Server roles
- Storage
- Network
- Database

Let's review each of these areas in the coming sections.

Core requirements

SharePoint Server 2019 has the following *minimum* hardware requirements:

Installation Scenario	Memory	Processor	Hard Disk Space
Dedicated Role Server	16 GB	64-bit, 4 cores	80 GB for system drive 80 GB for each additional drive

When planning for a highly performant, highly available server deployment, it is recommended that you choose a MinRole topology with dedicated server roles and redundant components. Current information on the SharePoint Server requirements can be found on the requirements page at `https://docs.microsoft.com/en-us/sharepoint/install/hardware-and-software-requirements-2019`.

Server infrastructure

As previously stated, most modern data centers utilize a high amount of virtualization. Virtualization, in and of itself, is a mechanism that allows a server (or *host*) to run more than one instance of a particular type of software (whether it's a full operating system running as a virtual machine, such as Windows Server 2019, or a container system, such as Docker, running applications).

In either of these cases, virtualization requires some overhead to allow the virtualization software to coordinate resources for the virtual machines or containers. This overhead subtracts from the overall amount of CPU cycles available to a service application and user requests.

With that being said, most SharePoint deployments will utilize virtual machines for some or all of the functions. Many organizations still choose to run database servers (due to their high transaction rates) in a more traditional fashion. If your organization chooses a virtualization strategy to support a SharePoint system, you will need to plan accordingly to ensure that the virtual infrastructure provided can meet the performance and user experience requirements of your organization.

Processors

Processors perform the computational tasks necessary to render content and applications. Per the minimum requirements, Microsoft recommends (four) 64-bit processor cores for a SharePoint Server instance. Task and context switching has a high degree of impact on overall performance. If your organization chooses to use virtualization, be careful to not oversubscribe or overcommit the number of processing resources presented to virtual machines.

Server roles

In the *Selecting and configuring a Farm topology* section of this chapter, we discussed the different roles available, particularly **Dedicated** versus **Shared**.

Shared roles, by their nature, take up more system resources than dedicated roles. Shared roles also introduce the potential for more task switching (time that the computer spends switching between threads to process transactions). For the highest-performing farm, it is recommended that each server performs only one role or function in a SharePoint farm. SharePoint services can be scaled both horizontally and vertically to accommodate multiple performance scenarios.

Storage

Storage (as a physical resource) is arguably one of the most important parts of a high-performance solution. Nearly all content and data stored or accessed inside a SharePoint environment is stored in some sort of database, which in turn rests on physical storage media. Storage is important to a SharePoint environment in two ways: the amount of data volume (capacity) and the speed at which data can be stored or retrieved.

When determining storage requirements for a high-performance system, there are a number of items to consider:

- The number of transactions (such as uploading, downloading, viewing, and so on), each of which generates I/O activity and is measured in **I/O Operations Per Second (IOPS)**
- The number of documents or other artifacts stored (and their associated metadata)
- The amount of data volume, including versioning

The general rule of thumb is to plan for I/O requirements first. A system's storage performance is generally governed by how many individual disks can be tasked at a particular time to act on data.

An individual disk's IOPS can be calculated using the following formula:

1000/(average disk latency + average disk seek time)

For example, a high-end 15,000 RPM mechanical drive with a seek time of 2.5 ms and a latency of 1.7 ms results in an IOPS value of 238. Typical mechanical hard drives deliver an IOPS ranging from 55 to 250, while SSD drive performance delivers 3,000–40,000 IOPS per disk.

Utilizing the principles you learned earlier for fault-tolerant design, you need to make sure that you design your storage in a way that can withstand the failure of one or more disk components. Disk subsystems typically achieve redundancy and performance using a mixture of technologies and configurations:

- RAID
- Multipathing
- Disk configurations
- Database capacity

Let's examine how each of these affects the storage infrastructure.

RAID

A RAID solution is a storage architecture that combines various aspects of performance, fault tolerance, and storage management. It is a form of storage virtualization, combining all or parts of physical disks to present a contiguous, logical allocation of storage to a host.

A group of disks operating in a RAID configuration is known as a **RAID set**. RAID subsystems work by saving data in either a striped (split into chunks and distributed) or mirrored (duplicated) fashion to the disks in the RAID set. Different combinations of striping and mirroring are referred to as **RAID levels**.

Additionally, RAID sets can include a feature called **parity**, which is a mathematical computation performed on data as a sort of error checking mechanism. In the event of disk failure, parity can be used to recover or rebuild the data on the inoperable disk. A RAID set is set to be in a **degraded** state when one or more disks in the set have failed but the volume as a whole is able to continue servicing requests.

To a large extent, a particular RAID set's performance depends not only on the RAID level but also on the read/write ratio. The read performance of a RAID set can be calculated using the formula. Each level has its own set of performance and redundancy characteristics, the most common of which are displayed in the following list:

- **RAID 0**: A RAID 0 configuration offers no fault-tolerance benefit. Data is striped between two or more disks evenly. In the event that any single disk in a RAID 0 set fails, all of the data is lost. RAID 0 provides a high level of overall read and write performance since it can engage multiple disks simultaneously to store and retrieve data. RAID 0 also provides the most possible storage, since there is no capacity overhead used to provide redundancy:

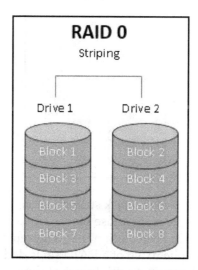

A RAID 0 set's performance can be expressed by simply multiplying the number of IOPS per disk and the number of disks—for example, a four-disk RAID 0 set with disks that can each perform 125 IOPS results in a max throughput of 1,000 IOPS.

- **RAID 1**: RAID 1 is the exact opposite of RAID 0. While RAID 0 has *no* fault tolerance whatsoever, RAID 1 can sustain the failures of up to half of the disks in a set. RAID 1 is commonly known as **drive mirroring** because incoming data is simultaneously written to two (or more) disks. RAID 1 only works with even numbers of disks. A two-disk RAID 1 set has a 50% capacity overhead as each disk has a full copy of all of the data. A two-disk RAID 1 set also has a 50% performance reduction on write activities since the data must be written twice. This configuration is commonly used for operating system drives:

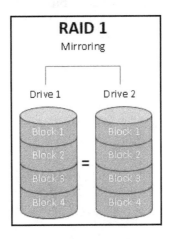

- **RAID 5**: A RAID 5 set utilizes disk striping (dividing a piece of data into multiple blocks and spreading those blocks among multiple drives) as well as parity (an error-checking calculation that is stored as a block on a drive). Effectively speaking, one drive of a RAID set is dedicated to storing parity calculations. RAID 5 requires a minimum of three drives to implement (which means a 33% storage capacity overhead). The most common configuration utilizes four drives, resulting in a 25% storage capacity overhead. RAID 5 volumes have very fast read times as multiple disks are being engaged. However, they do have a bit of performance overhead when both computing parity for writing as well as computing parity on the fly when the set is in a degraded state. This configuration is commonly used for file shares or database volumes with low to medium activity:

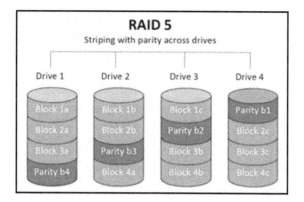

- **RAID 6**: Building on the parity concept of RAID 5, RAID 6 introduces an additional parity bit to allow a drive array to sustain failures on two drives. This increased parity leads to both a reduction in storage capacity as well as increased performance overhead since parity is computed and written twice. RAID 6 volumes require a minimum of four disks to implement. This configuration is commonly used for file shares or database volumes with low to medium activity:

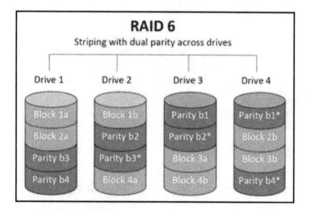

- **RAID 10**: RAID 10 is a combination of the capabilities of both RAID 1's mirroring and RAID 0's striping. It's commonly configured with four disks. RAID 10 has excellent read performance, as well as a 50% write performance penalty. RAID 10's total capacity is 50% of the disk capacity, but it can recover from up to two failed disks. This configuration provides both the highest performance and highest recoverability, but it also incurs the highest unusable amount of unusable storage—effectively, half the disks in the array. RAID 10, from a performance and resilience perspective, is typically the best choice for high-volume databases:

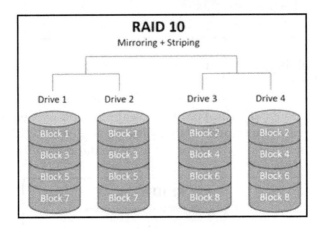

As you can see from the preceding diagrams and explanations, a RAID 10 disk subsystem is likely the best choice for high-performance SharePoint farms. Microsoft recommends the use of RAID 10 for SharePoint and SQL server configurations.

Multipathing

From a storage connectivity standpoint, multipathing allows a computer to connect to external storage through more than one route. Multipathing is typically configured when connecting to **Storage Area Networks** (**SANs**) via iSCSI or fiber storage networking. Configuring multipathing requires either multiple single-port storage adapters or a single multi-port storage adapter and a multipathing software package.

If your SharePoint farm or SQL database server environment uses physical servers (as opposed to virtualized servers), multipathing is an excellent way to provide both storage redundancy and storage performance.

Disk configurations

If you are using virtual machines in your SharePoint environment, you should evaluate the way you attach them to physical storage, including the following:

- Disk connection methods
- Fixed versus dynamic or thin-provisioned volumes

When you connect or attach disks to a virtual machine and begin configuring it, you generally have three options: exposing raw disk storage (in Hyper-V, this is referred to as a pass-through disk) to the virtual machine, connecting to raw disk using iSCSI or using **Network Attached Storage (NAS)**), or provisioning a VHD/VHDX file on a storage array and attaching that to the virtual machine configuration.

In earlier versions of Hyper-V, pass-through disks were generally considered the fastest method because there were fewer layers between the virtual machine and the disk access. Modern Hyper-V boasts a negligible difference, so it is probably best not to use them as they could present challenges to high availability or snapshots. You may want to test the performance between direct attach storage methods (such as iSCSI) to see whether they yield any performance gain.

If you decided to provision standard VHD/VHDX virtual hard drive files, it's important to provision them as fixed disks (where the entire volume of the virtual disk is pre-allocated) as opposed to using dynamically expanding disks (sometimes referred to as thin provisioning). Using thin provisioning or dynamically expanding disks greatly reduces the performance of a disk's volume since the storage is no longer contiguous. Disks serving the I/O requests will likely have to travel more to read or write data, introducing latency into the equation.

Database capacity

Part of the storage planning calculation is knowing (or estimating) how much data is stored. When calculating capacity needs, the following formula can be used:

Content database size = $((D \times V) \times S) + (10\ KB \times (L + (V \times D)))$

Here, we have the following:

- *D* is the number of documents to be stored, including any growth (most organizations tend to estimate growth over a period of 3 or 5 years). For computing departmental or project sites, you may want to use a number such as 10 or 20 documents per user; for My Sites, you may want to use a larger number (such as 500 or 1,000 per person).
- *V* is the number of versions that are stored (must be non-zero).
- *S* represents the average size of a document. Depending on the sites that will be stored in a particular content database, it may be useful to re-estimate this value for each content database (for example, My Sites will likely have a different usage profile than a departmental or project site).
- *L* is the estimated number of list items per site. If you have no pre-existing estimates, Microsoft recommends using a value of three times the amount of *D*.
- 10 KB represents the amount of storage volume consumed by metadata. If you plan on deploying a very metadata-driven environment, you may wish to adjust this number.

By applying the preceding formula to the following values, you can calculate the amount of storage capacity that a particular content database might be projected to need:

Variable	Value
Number of documents (*D*)	500,000
Number of versions per document (*V*)	5
Number of list items (*L*)	1,500,000
Average size per document	50 KB

The calculation is as follows:

*Content database size = (((500,000 x 5)) x 50) + ((10KB * (1,500,000 + (500,000 x 5)))*, or 165 GB of total storage.

Network

In order to produce a high-performance farm, you also need to ensure that the network connectivity between servers is optimal. Generally speaking, you should adhere to these guidelines:

- All servers should have a **Local Area Network (LAN)** bandwidth and latency to SQL servers used in the SharePoint farm. The latency should be 1 ms or less.
- Microsoft does not recommend connecting to a SQL environment over a **Wide Area Network (WAN)**.
- If configuring a site-redundant solution with SQL AlwaysOn located in a remote data center, you should ensure that you have adequate WAN bandwidth to perform the necessary log shipping or mirroring activities.
- Web and application servers should be configured with multiple network adapters to support user traffic, SQL traffic, and backup traffic.
- If connecting to iSCSI storage, ensure that any iSCSI network adapters are configured to only be used for iSCSI communication and not normal network communication.

These recommendations will help you build a farm that is both resilient and high-performing.

Summary

Designing a performant, reliable, SharePoint environment takes a lot of planning. Some of the areas that we touched on in this chapter included resilient and fault-tolerant design principles, using MinRole to manage server role deployments, and disaster recovery planning and scenarios.

Accessibility is also an important factor to consider when architecting an environment. We reviewed how localization can help make SharePoint content accessible to users with different language needs. Security is a large part of any modern system design; we discussed how Azure Information Protection and Rights Management services can be used to manage access and control the actions that others can perform against documents.

SharePoint Server 2019 is a capable content management and application platform and can be integrated with numerous Office 365 workloads and provide user experiences that cross platforms. This ability gives users and organizations the best of each platform's capabilities.

Finally, you learned about some of the steps you can take to design a high-performing farm, including processor, memory, storage, and networking recommendations.

In the next chapter, you will learn how to effectively manage and maintain a SharePoint farm environment, involving tasks such as configuring service applications and managing storage.

3
Managing and Maintaining a SharePoint Farm

In Chapter 2, *Planning a SharePoint Farm*, you learned how to plan and design SharePoint Server 2016- and 2019-based farms. Selecting a farm topology using MinRole will help you lay the foundation for a successful deployment. However, that's just the beginning.

Up to this point, we've discussed the concepts of farms, roles, sites, and service applications, but haven't really seen them up-close or gone into their configuration. In this chapter, however, we're going to start configuring some components to make the SharePoint Server farm useful.

In this chapter, we're going to cover the following topics:

- Configuring and maintaining core infrastructure to support a SharePoint deployment
- Deploying and configuring service and web applications
- Planning and configuring OneDrive/My Site access
- Planning and configuring user profiles
- Monitoring
- Updating SharePoint Server and validating the installation
- Selecting an upgrade path
- Managing SharePoint workflows
- Configuring a content type hub
- Troubleshooting performance issues
- Configuring SMTP authentication for a SharePoint farm

This is a lot of ground to cover, so let's get started!

Configuring and maintaining core infrastructure to support a SharePoint deployment

From an infrastructure perspective, Microsoft's design principles primarily focused on using MinRole-based farm topologies. MinRole is new to SharePoint Server 2016 and later. If you're not quite sure what MinRole is, please review Chapter 2, *Planning a SharePoint Farm*, for information on MinRole server roles and what services comprise each role.

When discussing the term *core infrastructure* as it relates to SharePoint Server, we're going to look at a few key areas:

- Server management from Central Administration
- SQL Server
- Distributed Cache

Let's take a look.

Server management from Central Administration

SharePoint configuration and management tasks are initiated within the Central Administration site. To easily access it, click on the **Central Administration** link from the **Start** menu on a server with SharePoint installed. You can also access it from any device by navigating to a server hosting the Central Administration site with a web browser.

 The SharePoint Central Administration site is typically configured to run on a custom port that's designated while you're configuring the SharePoint Server installation wizard. If you don't remember the port configuration or want to change it, launch the SharePoint Products Configuration Wizard to update the Central Administration setting or use the Set-SPCentralAdministration PowerShell cmdlet. As a reminder, with SharePoint Server 2016 and later, by default, SharePoint Central Administration is only provisioned on the first server in a farm where the SharePoint Products Configuration Wizard is run. It can be installed on additional servers at a later date using the SharePoint Products Configuration Wizard, PowerShell, or the Central Administration user interface on an existing server.

After launching the **Central Administration** site, you'll see the configuration, management, and monitoring options that are available, as shown in the following screenshot:

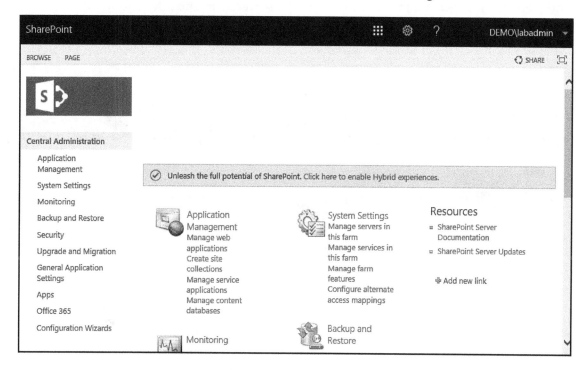

Under **System Settings**, there are four core setting areas to examine:

- **Manage servers in this farm**
- **Manage services in this farm**
- **Manage farm features**
- **Configure alternate access mappings**

Let's briefly look at each of those areas.

Manage servers in this farm

This page displays general information about the server farm, including information about the configuration database, servers that are members of the farm, and what role configuration they are using:

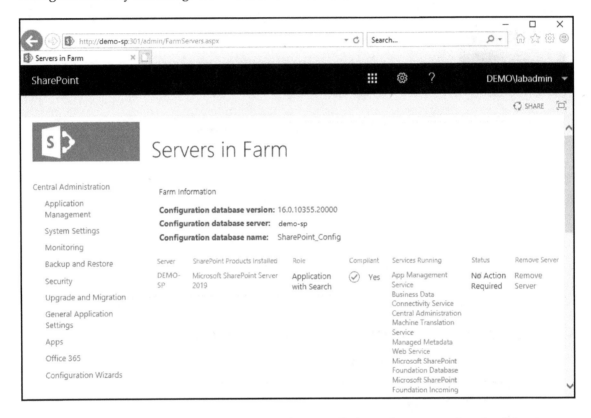

Clicking on the server name in the **Server** column will show the status of each of the services on the server. Clicking **Remove Server** will guide you through the process of removing a server from the farm.

Manage services in this farm

Selecting the **Manage services in this farm** option will display a page listing all the services across the farm, as shown in the following screenshot:

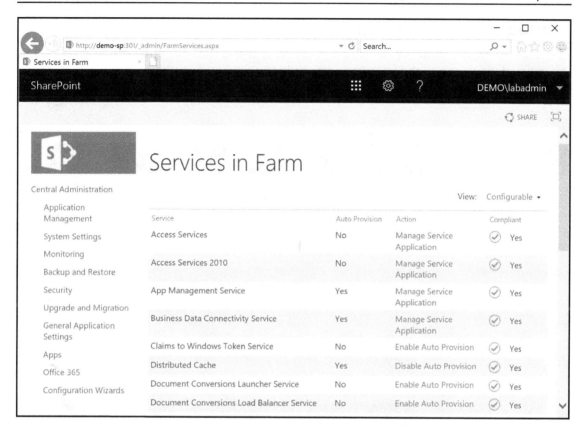

The **Auto Provision** column shows the current configuration status of the service across the farm. If a service has **Yes** in this column, this means that service instances for this role will be enabled and started on the appropriate MinRole-enabled servers in the farm. If a service has **No** under the **Auto Provision** column, then the service will be disabled on MinRole-enabled servers in the farm.

The **Action** column displays tasks that can be performed for a given service. A service's auto-provisioning status can be enabled or disabled via this column. Additionally, services that support service applications can be managed from here.

The **Compliant** column displays whether the service matches the expected configuration across all the servers in the farm. If a server's existing configuration is out of compliance, a link to fix it will be displayed.

Manage farm features

The **Manage Farm Features** page displays SharePoint-wide features that can be activated or deactivated, as shown in the following screenshot:

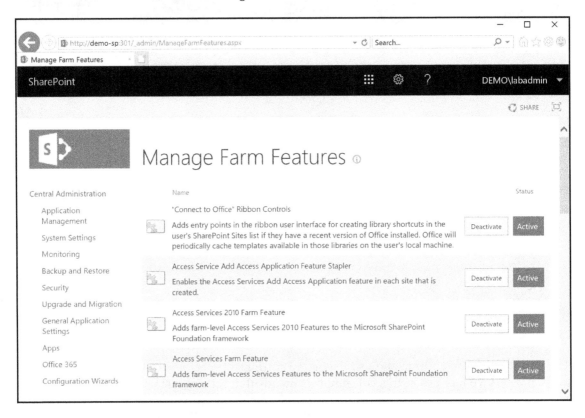

The following features can be enabled or disabled on this page:

"Connect to Office" Ribbon Controls	Access Service Add Access Application Feature Stapler
Access Services 2010 Farm Feature	Access Services Farm Feature
Cloud Video Thumbnail Provider	Data Connection Library
Excel Services Application View Farm Feature	Excel Services Application Web Part Farm Feature
Farm Level Exchange Tasks Sync	Global Web Parts
Office.com Entry Points from SharePoint	Office Synchronization for External Lists

SharePoint Server to Server Authentication	Site Mailboxes
Social Tags and Note Board Ribbon Controls	Spell Checking
User Profile User Settings Provider	Visio Process Repository
Visio Web Access	

In later chapters, we'll revisit this page to enable additional farm services.

Configure alternate access mappings

Alternate access mappings are used to direct or redirect users to the correct URLs for sites and applications. Alternate access mappings allow administrators to segment traffic into *zones*, based on the incoming URL. The configuration page for **Alternate Access Mappings** is shown in the following screenshot:

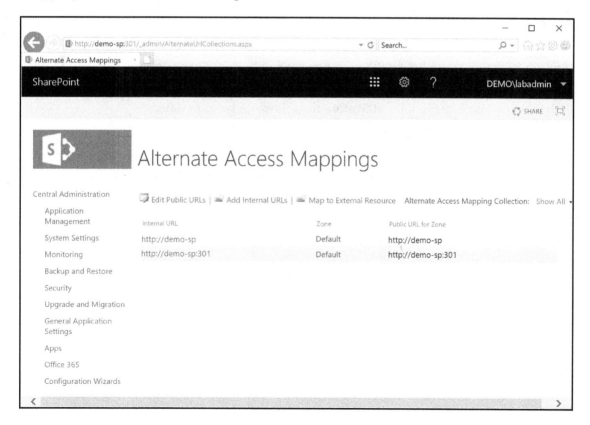

For example, you may have a SharePoint site called *Products* that lists your organization's products. This site could be available to both internal and external users. Your internal users may reach the *Products* site by browsing to `http://server/products`, while the extranet or partner organization users may reach the same content by going to `http://www.yourcompany.com/products`. One of the reasons you might use an alternate access mapping is to use different authentication providers to check and grant access to the site.

 The alternate access mappings feature is still available but has been deprecated in favor of a new process called *host-named site collections*. We'll configure host-named site collections later in this chapter.

The **Configure alternate access mappings** feature is still available in both SharePoint Server 2016 and 2019 but has been de-emphasized.

SQL Server

In Chapter 2, *Planning a SharePoint Farm*, we reviewed some of the database server requirements and recommendations for planning SQL Server-related tasks for SharePoint Server.

SQL Server is used by SharePoint to store the following types of data:

- Server and farm configuration
- Content such as document libraries and files, list data, web part information, user profile information, or data related to SharePoint web applications
- Service application data

Databases used to support SharePoint Server farms are typically created when setting up the SharePoint Products Configuration Wizard. Microsoft does not support directly querying or modifying SharePoint Server databases. Only SharePoint applications and supported APIs should be used to communicate with databases.

Distributed Cache

Introduced in SharePoint Server 2013, the Distributed Cache service provides in-memory caching services for many farm features. The Distributed Cache service is built on Windows Server AppFabric. It's important to note the following about the Distributed Cache service:

- The supporting service – that is, **AppFabric Caching Service** – should not be administered directly in the **Services** control panel.
- You should never launch the applications located in the **AppFabric for Windows Server** folder unless instructed to do so by Microsoft support.
- You should not attempt to implement additional security controls for AppFabric.

The following SharePoint Server features depend on the Distributed Cache service:

- Microblog Features and Feeds
- Login Token Cache
- Activity Feed Cache
- Activity Last Modified Time Cache
- OneNote Throttling
- Access Cache
- Search Query Web Part
- Security Trimming Cache
- App Access Token Cache
- View State Cache
- Default Cache

There are some administrative tasks you may need to perform with the Distributed Cache service:

- Stopping or starting the Distributed Cache service
- Adding or removing a server in a Distributed Cache cluster
- Changing the memory allocation for the Distributed Cache service
- Changing the service account of the AppFabric caching service
- Performance tuning
- Repairing a cache host

We'll look at the supported methods for each of these tasks in the following subsections.

Stopping or starting the Distributed Cache service

Distributed Management may need to be stopped or managed during upgrades. The Cache service can be stopped or started using **Central Administration**. To do this, follow these steps:

1. Launch **Central Administration** and select **Application Management**.
2. From **Service Applications**, click **Manage Services on Server**.
3. On the **Services on Server** page, find the Distributed Cache service.
4. Under the **Action** column, click **Stop** to stop the Distributed Cache service (if it is running) or click **Start** to start the service (if it is not currently running).

However, if the cache host is a member of a Distributed Cache cluster, use the process detailed in the *Adding or removing a server in a Distributed Cache cluster* section.

Adding or removing a server in a Distributed Cache cluster

If you need to perform maintenance activities on a server that is part of a Distributed Cache cluster, you can stop or start the Distributed Cache service using the SharePoint Management Shell. Follow these steps:

1. Launch the **SharePoint Management** shell on the Distributed Cache cluster node you wish to administer.
2. Run `Add-SPDistributedCacheServiceInstance` to add the server to the cache cluster and start the service. Alternatively, use `Remove-SPDistributedCacheServiceInstance` to remove the server from the cluster and stop the service.

Removing a Distributed Cache cluster host unregisters the service from the server, so it will no longer appear. Adding the server back to a cluster will re-register the service.

Changing the memory allocation for the Distributed Cache service

When configuring SharePoint Server, the Distributed Cache service is automatically allocated 10% of the server's physical memory. If you install or remove memory in a Distributed Cache host, or if the server is configured to be a dedicated Distributed Cache server, you may need to update the configuration for the Distributed Cache service.

When calculating the amount of memory to allocate, use this formula:

((Total Physical Memory - 2GB) / 2) * 1024

For example, if your server has 24 GB of physical memory, we get the following:

((24 - 2) / 2) * 1024 = 11,264 (11 GB)

A maximum of 16 GB can be allocated to the Distributed Cache service on a cache host.

Viewing cache configuration

To view the existing configuration, follow these steps:

1. Launch SharePoint Management Shell on a server configured with the Distributed Cache role.
2. Run the following cmdlets:

```
Use-CacheCluster
Get-AFCacheHostConfiguration -ComputerName $Env:ComputerName -
CachePort "22233"
```

The console will display the current cache configuration.

Configuring the cache

If you change the amount of memory that's installed in a server, you will need to update the cache configuration to take advantage of it. To update the configuration, follow these steps:

1. Launch **Central Administration**.
2. Stop the Distributed Cache service on all hosts by selecting **Services on Server** and then **Stop** under the actions for the Distributed Cache service.
3. Launch the **SharePoint Management** Shell on a server configured with the **Distributed Cache** role.

4. Run the following cmdlet (where `<value>` is the size in megabytes):

```
Update-SPDistributedCacheSize -CacheSizeInMB <value>
```

5. From **Central Administration**, select **Services on Server** and then **Start** under the actions for the Distributed Cache service.

The configuration cmdlet only needs to be run once. All the servers will use the same setting.

Changing the service account of the AppFabric caching service

When a SharePoint Server farm is initially configured, the SharePoint Products Configuration Wizard sets the service account for many services, including the AppFabric Caching service, to the farm service account. Due to the security requirements of your organization, it may be necessary to change this to a different account (such as a SharePoint managed account). To do this, follow these steps:

1. On an **Active Directory Domain Controller** (or a computer with **Remote Server Administration Tools** installed), launch **Active Directory Users and Computers**.
2. From **Active Directory Users and Computers**, create a new domain user account.
3. Launch **SharePoint Management Shell**.
4. Create a new **SharePoint Managed Account** using the following process (entering the `DOMAIN\username` and `password` details of the account you created in *step 1*):

```
$Credential = Get-Credential
New-SPManagedAccount -Credential $Credential
```

5. Set the **Distributed Cache** service so that it uses the new **SharePoint Managed Account**:

```
$Farm = Get-SPFarm
$CacheService = $Farm.Services | ? {$_.Name -eq
"AppFabricCachingService"}
$ManagedAcount = Get-SPManagedAccount -Identity
$Credential.UserName
$cacheService.ProcessIdentity.CurrentIdentityType = "SpecificUser"
$cacheService.ProcessIdentity.ManagedAccount = $ManagedAccount
$cacheService.ProcessIdentity.Update()
$cacheService.ProcessIdentity.Deploy()
```

The Distributed Cache service will be updated to use the new account. In farms where there is more than one Distributed Cache server, stop the Distributed Cache service on all other Distributed Cache hosts.

Performance tuning

While many SharePoint administrators don't update the cache settings, it is important to know that they can be viewed and modified using the `Get-SPDistributedCacheClientSetting` and `Set-SPDistributedCacheClientSetting` cmdlets, respectively. For more information on specific performance tuning values, see `https://docs.microsoft.com/en-us/sharepoint/administration/manage-the-distributed-cache-service#fine-tune-the-distributed-cache-service-by-using-a-powershell-script`.

Repairing a cache host

At some point, a Distributed Cache host may enter a non-function or non-responsive state and its health status will trigger an alert in the Health Rules in Central Administration. If this happens, it may be necessary to remove and restore the Distributed Cache Host. To do so, log in to the non-functioning Distributed Cache host and follow these steps:

1. Launch an elevated SharePoint Management Console prompt.
2. Run the `Remove-SPDistributedCacheServiceInstance` cmdlet.
3. Run the `Add-SPDistributedCacheServiceInstance` cmdlet.

For more detailed information on configuring the Distributed Cache service, see `https://docs.microsoft.com/en-us/sharepoint/administration/manage-the-distributed-cache-service`.

Now that you've seen where to administer SharePoint Server at a high level, we'll start configuring service applications.

Deploying and configuring service and web applications

In SharePoint Server terminology, web applications are the constructs that are used to provide the configuration and hosting infrastructure for SharePoint site collections. Each web application starts with an **Internet Information Services (IIS)** website. Each individual web application is connected to its own IIS website (using either a unique or a shared application pool). Frequently, each IIS site is configured with a unique domain namespace, which can be used for both application structuring and topology, as well as a security measure to mitigate cross-site scripting attacks.

Each web application has at least one content database associated with it. SharePoint Server 2016 and SharePoint Server 2019 both support up to 20 web applications per farm. Since each web application has an IIS site associated with it, you can only have one (by default) that answers on port 80 or 443. In previous versions of SharePoint Server, you could accomplish hosting multiple applications on a single port by using host headers or alternate access mappings.

While these methods are still supported, Microsoft currently recommends using *host-named site collections* moving forward unless absolutely necessary, as this is the underlying technology being deployed and supported in SharePoint Online. Host-named site collections allow you to assign unique DNS names to site collections (such as http://intranet.fabrikam.com, where intranet is the hostname assigned to the site collection), as opposed to using a longer path to the site collection (such as http://fabrikam.com/sites/intranet).

 A host header is *different* to a host-named site collection. A host header is an IIS property and is handled by the underlying web service. Host headers are to be used with path-based site collections *only*. With host-named site collections, SharePoint performs site name resolution itself. Configuring a host header will interfere with the name-to-site resolution.

For host-named-based site collections, Microsoft currently recommends an architecture similar to the following as a best practice:

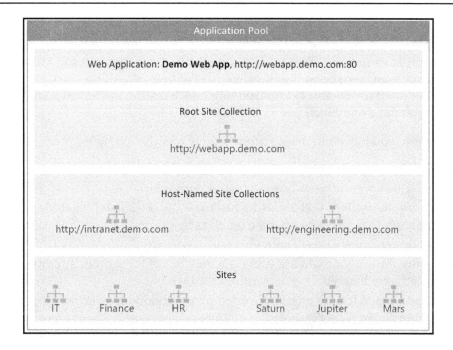

While web applications are generally used to provide a framework for users to interact with data, service applications are used to share features and capabilities across web applications or farms. Examples of service applications include the following:

- Managed Metadata
- Secure Store
- PerformancePoint
- User Profile

These capabilities allow developers and administrators to leverage common data and configurations across a farm or organization.

 For an example of creating and configuring a service application, take a look at Chapter 7, *Planning and Configuring Managed Metadata*.

We'll examine tasks for creating and configuring a web application next.

Creating a web application

While Microsoft recommends creating host-named site collections moving forward (it's the technology used to support SharePoint Online), at the time of writing, the Central Administration interface does not support doing this. Creating host-named site collections requires that you use PowerShell.

Before you begin, you'll need to make sure you have met the following prerequisites:

- Ensure the account you are using has the securityadmin fixed server role on the SQL Server instance hosting SharePoint.
- Ensure the account you are using has the db_owner fixed database role on all databases that are to be updated (such as the master and SharePoint_Config databases).
- Ensure the account you are using is a member of the Administrators group on the server where you are performing the task.
- Select a name for the web application (in the following example, we're going to use the name Demo Web App).
- Create or select an existing **SharePoint Managed Account** (you can refer to the steps provided in Chapter 2, *Planning a SharePoint Farm*, for configuring a managed account).
- Ensure no existing website or application is listening on the port specified.

Once you have met these prerequisites, you can follow these steps to create a basic web application that will be used to host your site collection. We'll be using the previous reference architecture diagram as a guide:

1. Launch **SharePoint Management Shell**.
2. Run the following command (replacing the parameter values with ones valid for your environment) to create the web application:

```
$WebAppPparams = @{
Name = 'Demo Web App'
Port = 80
ApplicationPool = 'DemoAppPool'
ApplicationPoolAccount = (Get-SPManagedAccount
"Demo\SPFarmServices")
AuthenticationProvider = (New-SPAuthenticationProvider -
UseWindowsIntegratedAuthentication)
}
New-SPWebApplication @WebAppParams
```

If you have an existing application using port 80 on this server, you'll receive an error when configuring the web application and will need to either change the port, add an interface to the server, or remove the existing configuration.

Once completed, the web application should be visible in the Central Administration interface when you navigate to **Central Administration | Application Management | Manage Web Applications**:

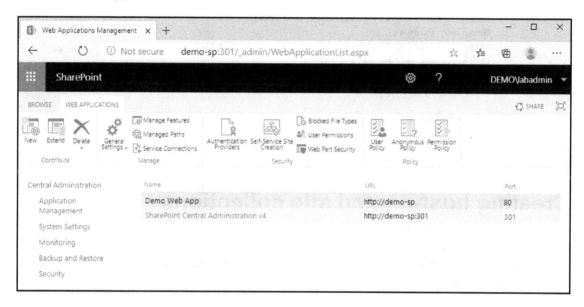

Once the web application has been created, you can start creating the site collections.

Creating a root site collection

Creating the root site collection will allow you to create sites with unique names. A root site collection is required for any deployment, as well as crawling content. The root site collection must have the same URL as the web application. At the time of writing, SharePoint prevents you from creating a host-named site collection with the same URL as a web application, so you'll need to create it as a path-based site collection.

Using the previously created web application configuration, we're going to create the root site collection:

1. Launch **SharePoint Management Shell**.
2. Run the following command (replacing the parameter values with ones that are valid for your environment) to create the root site collection:

```
$SiteParams = @{
Url = "http://$($env:COMPUTERNAME)"
Name = "Portal"
Description = "Portal root site collection"
OwnerAlias = "demo\labadmin"
Language = "1033" # English
Template = "STS#0" # Team site template
}
New-SPSite @SiteParams
```

After a few moments, the site collection should be created.

Creating host-named site collections

Now that all of the underlying application infrastructures have been completed, you can create the host-named site collections. As we mentioned previously, at the time of writing, host-named site collections cannot be created in the **Central Administration** user interface, although path-based configurations are supported.

New host-named site collections are created with the New-SPSite - HostHeaderWebApplication syntax. Be careful, though, as the cmdlet name is a little misleading. Host headers typically refer to a web server configuration, but you are not configuring IIS host headers.

When using the reference architecture as a guide, follow these steps to create host-named site collections:

1. In your organization's DNS console, add a **CNAME** record for each URL pointing to the hostname of the SharePoint farm. In our example, we're going to create host-named site collections for intranet.demo.com and engineering.demo.com, so we'll need **CNAME** records that map those names to the existing web application (webapp.demo.com).
2. Launch **SharePoint Management Shell**.

3. Run the following command (replacing the parameter values with ones valid for your environment) to create some host-named site collections:

```
$Sites =
@(@{Url="intranet.demo.com";Name="Intranet"},@{Url="engineering.dem
o.com";Name="Engineering")
$HostNamedSiteParams = @{
HostHeaderWebApplication = ((Get-SPWebApplication "Demo Web
App").Url)
OwnerAlias = "demo\labadmin"
Language = "1033" # English
Template = "STS#0" # Team site template
}
Foreach ($Site in $Sites) { New-SPSite @HostNamedSiteParams -Name
$Site.Name -Url "http://$($Site.Url)" }
```

Once the site collections have been created, you can navigate to them and add additional subsites and content as appropriate:

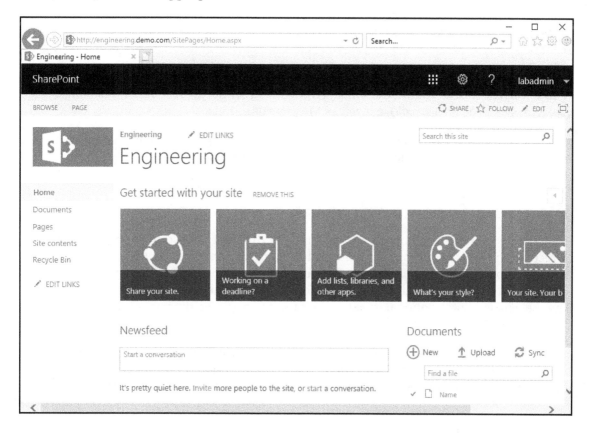

In the preceding example, the STS#0 (Team Site) template was used. You can select other templates as well. Additional template IDs are available by running `Get-SPWebTemplate` from the SharePoint Management Shell.

Now that you understand how web applications are created and used, we'll look at using them to host applications and services. Next, we will begin exploring the User Profile service.

Planning and configuring user profiles

The User Profile service is used by SharePoint Server to store data about users. It's required for My Sites and OneDrive, social networking features such as tagging and newsfeeds, creating and managing audiences and organizational charts, and delivering user profile information to other services or web applications (either on the local farm or connected farms). The User Profile service is also used by servers for accessing resources on behalf of users in server-to-server authentication scenarios, as well as in SharePoint Hybrid scenarios.

The User Profile Service has a dependency on the Managed Metadata service application (you'll learn how to configure the Managed Metadata service in Chapter 7, *Planning and Configuring Managed Metadata*), as well as My Site collections.

In this section, we'll look at an overview of some of the planning tasks for the User Profile service before configuring the User Profile service application.

Planning for user profiles

User profile data can be gathered from a variety of different data sources, including Active Directory, third-party business systems, scripting, or manual entry by end users. It's important to note that *user profiles* are different to *user accounts*. The SharePoint user account store is used to manage security and access controls to various parts of the platform. The *user profile* is informational in nature and used to organize data about the users and their relationships with other data entities.

In SharePoint Server 2016 and later, data is typically ingested into the User Profile service through a process called synchronization. You can import data into a user profile using Active Directory Domain Services Import or Microsoft Identity Manager 2016.

ForeFront Identity Manager (**FIM**) was included with previous versions of SharePoint Server to facilitate user profile synchronization. It is no longer included with SharePoint Server 2016 or 2019. You can still use an Identity Manager-based solution to synchronize user profile data into SharePoint Server 2016 and later, but it requires additional infrastructure and the newer Microsoft Identity Manager platform. Configuring Microsoft Identity Manager integration is outside the scope of this book (and generally involves the assistance of an identity specialist), but you can learn more about it here: `https://docs.microsoft.com/en-us/sharepoint/administration/microsoft-identity-manager-in-sharepoint-server-2016`

Planning for a successful User Profile service deployment involves doing the following:

- **Identifying what data will be collected**: This could include basic information such as name or job title, and organizational data such as a manager or team members, as well as more detailed data such as years in their current role or skills.

- **Identifying which properties will be public or private**: If the User Profile service application is used to store data of a more private nature (such as performance review metrics), you may want to flag that data as only being available to the user themselves.

- **Identifying what sources will be used to contribute data**: Most organizations will synchronize some data from Active Directory, though they may have other internal application sources (such as human resources or a payroll application) that may contain additional data points to be included.

- **Determining how profile properties will be used for any applications or personalization policies**: Mapping out the requirements of applications will be critical to knowing whether data *needs* to be included in the SharePoint User Profile service.

Each organization's requirements will be different in terms of balancing the amount of data collected with how it will be used and what will be visible to other people in the organization.

There are many profile features and settings to be explored and discussed with key stakeholders, administrators, and application developers. You can find a full list of policy and property settings here: `https://docs.microsoft.com/en-us/sharepoint/administration/plan-user-profiles`.

Next, we'll create the User Profile service application.

Creating a User Profile service application

If the User Profile service hasn't already been created in your SharePoint farm, follow these steps to build it:

1. Launch **Central Administration**.
2. Navigate to **Application Management**, and then select **Manage service applications** from the **Service Applications** section.
3. On the ribbon, from **Create grouping**, click **New** and select **User Profile Service Application**:

4. Configure the settings for the **User Profile Service** application:

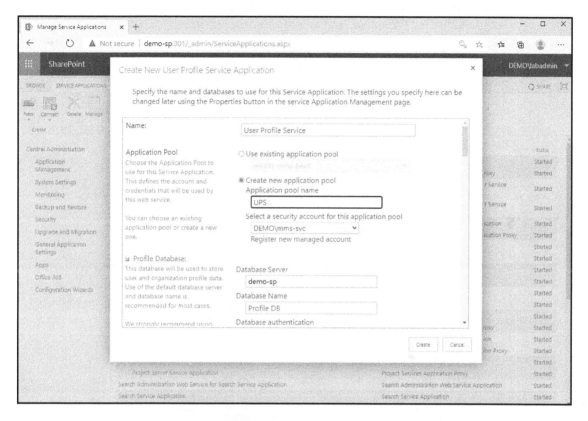

5. In the **Application Pool** section, either choose an existing application pool or select **Create new application pool** to create a new one.

6. In the **Profile Database** section, enter the name of the database server you want to host the User Profile service application database, enter a value for the name, and select an authentication type (Windows authentication is recommended). If using SQL Database Mirroring, configure the **Failover Server** option.

7. In the **Synchronization Database** section, enter the name of the database server you want to host the User Profile service application synchronization database, enter a value for the name, and select an authentication type (Windows authentication is recommended). If using SQL database mirroring, configure the **Failover Server** option.

8. In the **Social Tagging Database** section, enter the name of the database server you want to host the User Profile service application social tagging database, enter a value for the name, and select an authentication type (Windows authentication is recommended). If using SQL database mirroring, configure the **Failover Server** option:

   - Leave the **My Site** sections blank for now.
   - In the **Site Naming Format** section, select a method for naming new personal sites.
   - In the **Default Proxy Group** section, leave the default option selected.
   - In the **Yammer Integration** section, select whether you wish to use Yammer for social features. *This will require an Office 365 tenant with an active Yammer subscription.* For this exercise, select the **Use on-premise SharePoint social functionality** option.

9. Click **Create**.

Once the **User Profile Service** has been created, we can move on to customizing.

Extending user profiles

If your organization needs to keep track of data or information as part of the user profile that is not part of the default user profile dataset, you can customize the User Profile service with additional properties. These properties may be used for reporting, creating and managing audiences, or integration with other applications.

To add, delete, or modify a user profile property, follow these steps:

1. Launch **Central Administration**.
2. Click **Application Management**, and then select **Manage service applications**.
3. Click on the **User Profile** service application.
4. From the **People** section, select **Manage User Properties**:

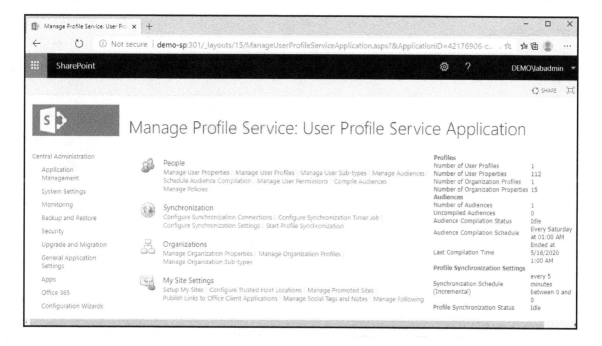

5. To add a property, click **New Property**. To edit or delete a property, select a property from the list and select **Edit** or **Delete**:

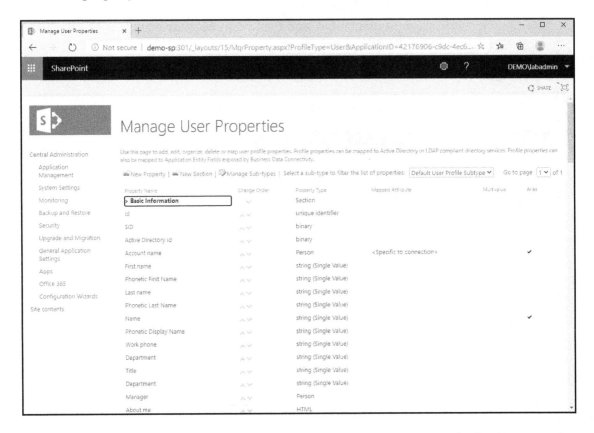

6. When adding a property, fill in the required values (**Name**, **Display Name**, and **Description**). If you have additional language packs installed, you can click **Edit Languages** to edit how properties are displayed to the end user. *Values for* **Name** *must comply with the URI schema name specification in RFC2396 (alphanumeric, start with a letter, can include symbols (., +, or -), and must not contain spaces):*

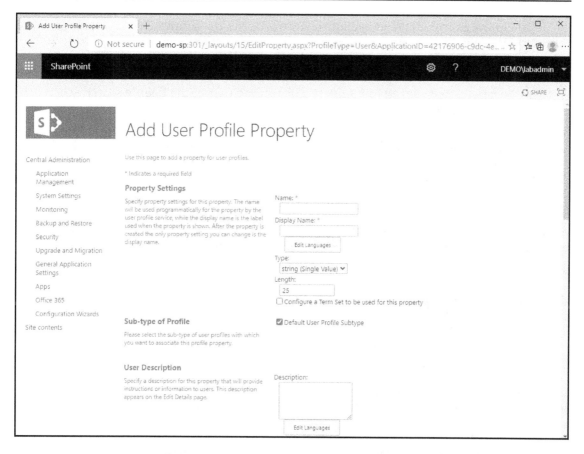

7. When adding a property, under **Property Settings**, select a **Type**. This type *cannot be changed* when you're editing; if you need to change the type of a property, you'll need to delete the property and start over.

8. In the **Policy Settings** section, select whether the policy is required and its privacy settings, as well as whether you want the user to be able to override the configuration.

9. In the **Display Settings** section, choose where the property is displayed (**User Profile** page, **Edit Details** page, or **Newsfeed**).

10. In the **Search Settings** section, choose whether the property should be indexed or aliased. Aliased properties are treated equivalent to their username when performing searching. The availability of the **Alias** checkbox depends on the privacy setting for the property being set to **Everyone** (public). If you mark a property as **Indexed** (default), it will become part of the **People** search scope schema and will be crawled.

11. If you have connected data sources (such as an HR application or a **Business Data Connectivity** (**BDC**) service), you can create a mapping between the data source and the User Profile service. (*Note: BDC sources only support Import, while other connected data sources also support Export*).

12. Click **OK** to finish creating the profile property.

At this point, you've learned how to configure the user profile service and extend it with additional properties. Next, we'll look at importing data from Active Directory.

Configuring user profile synchronization

The User Profile service, as mentioned previously, is used to gather information about the user. By default, it has no data associated with it. You'll need to connect one or more data sources to it in order to gather user data.

In this section, we'll walk through configuring **SharePoint Active Directory Import**, which is the most common method for getting user data into the User Profile service.

Configuring SharePoint to use Active Directory

The first step in this process is to update the User Profile service so that it uses Active Directory. Follow these steps:

1. Launch **Central Administration**.
2. Click **Application Management** and then select **Manage service applications**.
3. Click **User Profile Service** and then select **Manage** from the **Operations** group on the ribbon.

4. Under **Synchronization**, select **Configure Synchronization Settings**.
5. Select the **Use SharePoint Active Directory Import** radio button and click **OK**:

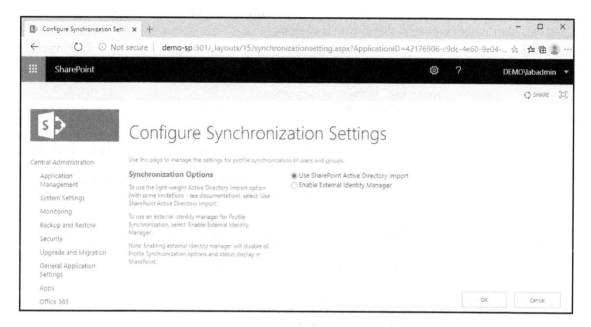

The User Profile service application will now use the native Active Directory import.

Creating a synchronization connection

At this point, the User Profile service has been prepared for Active Directory import. The next task is to connect the User Profile service to one or more Active Directory environments.

The core prerequisite for configuring synchronization is a service account. Configuring a synchronization connection requires an Active Directory user account with the **Replicating Directory Changes** permission delegated. Most SharePoint Server services *cannot* use a Managed Service Account or Group Managed Service Account and must use a normal user account.

Follow these steps to set up an account with the correct permissions and configure the synchronization connection:

1. On a server that has **Active Directory Remote Server Administration Tools**, launch an elevated PowerShell session.
2. Run the following command, substituting the appropriate values for the domain and user account:

```
$SamAccountName = "sps-adimport-svc"
$Name = "SharePoint Active Directory Import"
$Password =
([System.Web.Security.Membership]::GeneratePassword(15,2))
$SecurePassword = ConvertTo-SecureString -AsPlainText $Password -
Force
New-ADUser -DisplayName $Name -SamAccountName $SamAccountName -Name
$Name -AccountPassword $SecurePassword -Enabled $True
$RootDSE = Get-ADRootDSE
$DefaultNamingContext = $RootDSE.defaultNamingContext
$ConfigurationNamingContext = $RootDSE.configurationNamingContext
dsacls $DefaultNamingContext /G
"$($SamAccountName):CA;""Replicating Directory Changes"$RootDSE =
Get-ADRootDSE
$DefaultNamingContext = $RootDSE.defaultNamingContext
$ConfigurationNamingContext = $RootDSE.configurationNamingContext
dsacls $DefaultNamingContext /G
"$($SamAccountName):CA;""Replicating Directory Changes"
"$($env:USERDOMAIN)\$($SamAccountName)"
"$($Password)"
```

3. Save the username and password values for later.
4. Launch **Central Administration**.
5. Click **Application Management** and then select **Manage service applications**.
6. Click **User Profile Service** and then select **Manage** from the **Operations** group on the ribbon.
7. From **Synchronization**, select **Configure Synchronization Connections**.

8. Click **Create New Connection**:

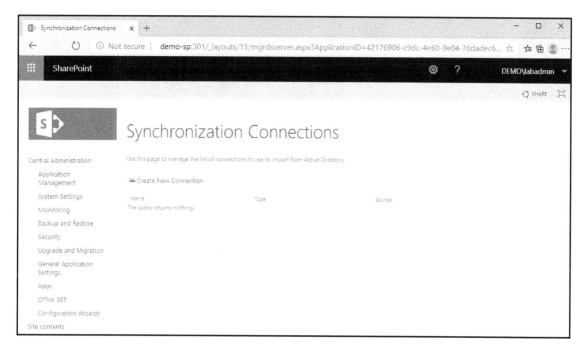

9. In the **Connection Name** section, enter a name that will uniquely identify this directory connection.

10. In **Type**, ensure **Active Directory Import** is selected.

11. In **Connection Settings**, enter the fully qualified domain name and set **Windows Authentication** as the authentication type.

12. Using the values saved in *step 3*, populate the **Account name** and **Password** fields:

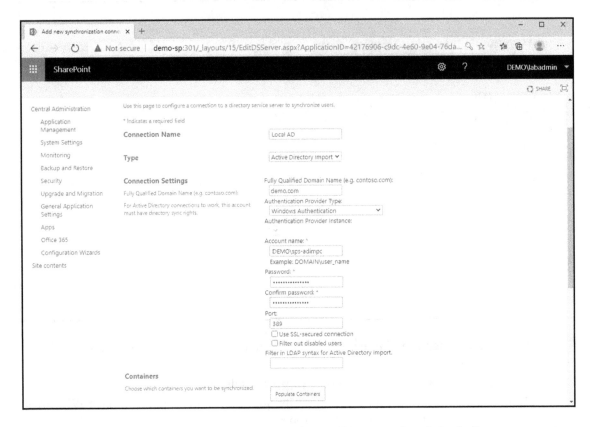

13. Ensure the **Port** value is set to 389. Active Directory, by default, listens on port 389, so you'll likely just leave this as-is. If you want to use SSL to perform synchronization, you will need to export the certificate from a domain controller and import it into the synchronization server before completing this task (alternatively, you can return to this connection later and update it).

14. Select filtering options. Here, you can choose whether to filter disabled accounts with the **Filter out disabled users** checkbox, as well as specify an LDAP query to narrowly scope objects.

15. Under **Containers**, select which organizational units to synchronize. Click the **Populate Containers** button to retrieve the organization units for the selected domain, and then click the selection box next to the name of the organizational unit to include it in the synchronization process:

16. Click **OK**.

With that, the synchronization connection should be configured and listed on the **Synchronization Connections** page.

Mapping user profile properties

By default, user profile properties are mapped to their corresponding AD attributes. However, there may be times when you need to update the default mappings or define mappings for custom properties.

To modify or update the mappings for user profile properties, follow these steps:

1. Launch **Central Administration**.
2. Click **Application Management** and then select **Manage service applications**.
3. Click **User Profile Service** and then select **Manage** from the **Operations** group on the ribbon.
4. Under **People**, select **Manage User Properties**.
5. Select a property to edit, and then click **Edit**.
6. Scroll down to **Property Mapping for Synchronization** to remove mappings. Alternatively, go to the **Add a New Mapping** section to add one.
7. Click **OK** when you're finished.

There are a handful of default mappings that are not displayed in the User Interface. These are as follows:

Active Directory attribute	User profile property
department	Department
displayName	PreferredName
dn	SPS-DistinguishedName
givenName	FirstName
mail	WorkEmail
manager	Manager
msDS-PhoneticDisplayName	SPS-PhoneticDisplayName
msDS-PhoneticFirstName	SPS-PhoneticFirstName
msDS-PhoneticLastName	SPS-PhoneticLastName
msDS-SourceObjectDN	SPS-SourceObjectDN
objectSid	SID
physicalDeliveryOfficeName	Office
proxyAddresses	SPS-ProxyAddresses
sAMAccountName	UserName
sn	LastName
telephoneNumber	WorkPhone
title	SPS-JobTitle
wWWHomePage	PublicSiteRedirect

If you have added custom properties or want additional data source mappings, edit the property mappings. Once you have made your edits, you'll need to synchronize the profiles.

Running a manual profile synchronization

The profile synchronization process is used to import data from connected directories. Follow these steps to initiate manual profile synchronization:

1. Launch **Central Administration**.
2. Click **Application Management** and then select **Manage service applications**.
3. Click **User Profile Service** and then select **Manage** from the **Operations** group on the ribbon.
4. Under **People**, select **Start Profile Synchronization**.
5. Select the **Start Full Synchronization** radio button and click **OK**:

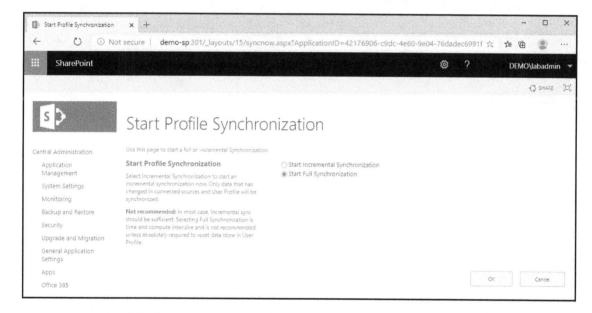

Most of the time, incremental synchronization is sufficient. However, if you add properties or update mappings, you'll need to run a full synchronization.

Once you have run a profile synchronization, you can view a user's profile properties to make sure data was imported successfully. To navigate to a user's profile, on the **Manage Profile Service** page, under the **People** section, select **Manage User Profiles**. From here, you can search for a profile and view its data:

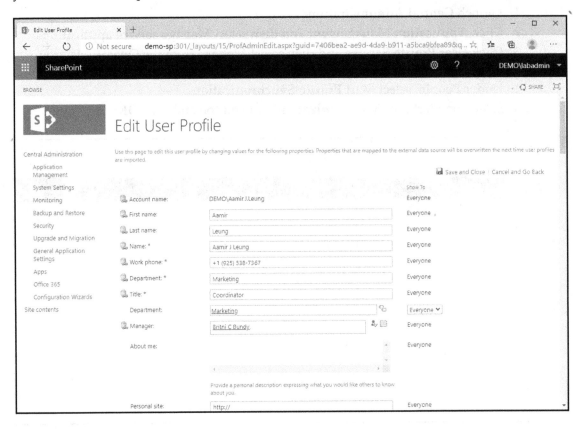

Once you have confirmed that user profile data has been successfully imported, you can configure a profile synchronization schedule.

Scheduling profile synchronization

As we learned in the previous section, profile synchronization imports data from the connected directory data source to the appropriate mapped attributes. Manual synchronization is just that – a manual, one-time process. To keep your User Profile service updated with the latest data, you'll want to configure a regular synchronization job.

To configure the synchronization schedule, follow these steps:

1. Launch **Central Administration**.
2. Click **Application Management** and then select **Manage service applications**.
3. Click **User Profile Service** and then select **Manage** from the **Operations** group on the ribbon.
4. Under **People**, select **Configure Synchronization Time Job**.
5. Select a time increment (minutes, hours, days, weeks, or months) and enter a numeric value:

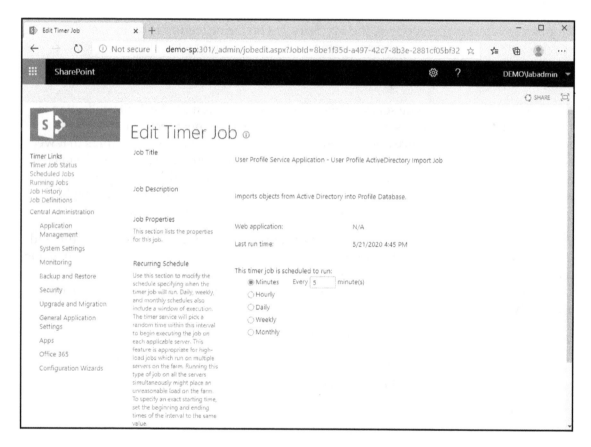

6. Scroll to the bottom of the page and click **OK**.

With that, the User Profile service application has been configured and is available to import and store data associated with users from a variety of sources. In the preceding section, it was configured to import from Active Directory.

Next, we'll begin planning and configuring My Site (also known as OneDrive).

Planning and configuring OneDrive/My Site access

From an architecture perspective, the **My Sites** or **OneDrive** feature is really just a specialized site collection that hosts individual user SharePoint sites. It has a specific template (the **My Site Host** template) that's used to store personal files and content. You can think of it as a potential replacement for standard file share-based home directories.

My Sites and the User Profile service have interdependencies: a My Sites site collection is required for the User Profile service (which we'll get to in the next section); My Sites requires a User Profile service application, while the User Profile service requires that we assign a My Sites collection as part of the configuration. My Sites also requires a Managed Metadata service application (you'll learn more about the Managed Metadata service application in `Chapter 7`, *Planning and Configuring Managed Metadata*). Configuring My Sites requires a little bit of back and forth with the User Profile service application.

A My Site host site collection can be created during the initial SharePoint Server setup and configuration. If it's been created, Microsoft recommends deleting that site collection, using a new web application dedicated to hosting My Sites, and then creating a new site collection in that web application. This provides the best option in terms of performance, scalability, and security.

In this section, we'll look at two common My Sites configuration tasks:

- Creating a My Sites collection
- Updating the User Profile service application

Let's get started!

Creating a My Sites collection

As the Microsoft recommendation is to create a separate web application for My Sites, we'll walk through the steps to do this here. The prerequisites for this process will be as follows:

- You need an additional IP address assigned to each server that will be hosting My Sites.
- A DNS; that is, a record representing each of the IP addresses that will be associated with My Sites is needed. In this example, we will add a DNS A record to map the name *my* to 10.0.0.6.

Once you have met these prerequisites, follow these steps to create the My Sites site collection and enable self-service provisioning for users:

1. Create a DNS A record for your My Sites host.
2. Launch **Central Administration**.
3. Click **Application Management** and then select **View all site collections**. If a My Sites collection has been configured, delete it.
4. In this example, we're going to host My Sites on a separate IP address on the same server so that we don't have to deal with host headers or alternate access mappings. You can use a script such as the following for each of your servers that will host My Sites, replacing the IP address and the DNS name for the host header in New-IISSite -BindingInformation:

   ```
   New-Item -Path C:\InetPub -Name MySites -ItemType Directory
   New-IISSite -Name "My Sites" -PhysicalPath "C:\InetPub\MySites" -
   BindingInformation "10.0.0.6:80:my.demo.com"
   ```

5. Create a new web application that will host the site collection. From **Central Administration**, click **Application Management** and then select **Manage web applications**.

6. From the ribbon, select **New**. Fill out the values of the web application. For this example, we'll use our newly created website, **My Sites**, by selecting it from the **IIS Web Site** section. If you don't have an SSL certificate already bound to the site, you won't be able to configure SSL, but you can come back and update the web application afterward:

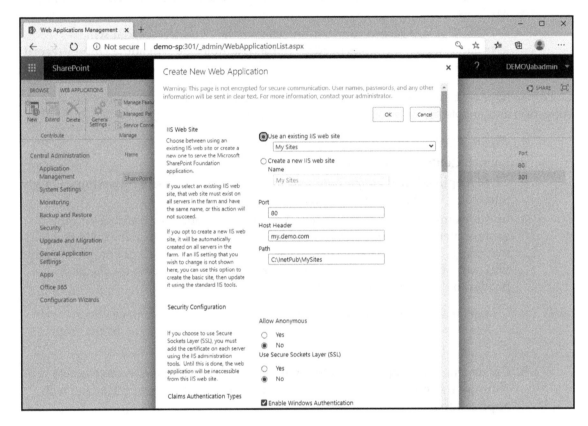

7. In the **Public URL** section, enter the DNS name you're going to assign to this **My Sites** instance.

8. In the **Application Pool** section, select the **Create new application pool** radio button, enter a name, and select a security account (like we did for the farm service account). You can create and register a new managed account from here as well:

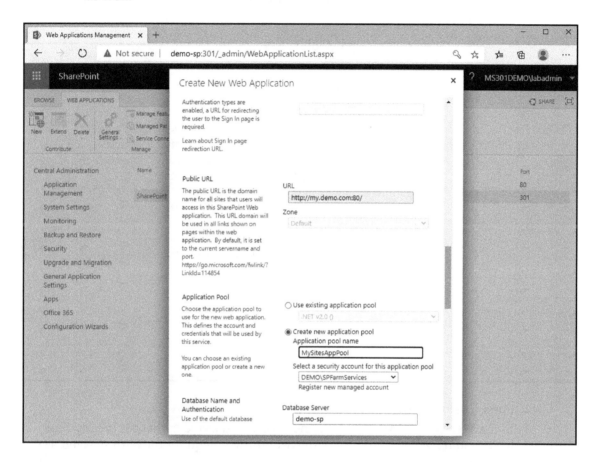

9. In the **Database Name and Authentication** section, enter the database server and database name you wish to use and choose an authentication type.
10. Scroll to the bottom of the page and click **OK**.
11. Once the web application has finished provisioning, click the **Create Site Collection** link:

Create New Web Application   ✕

The Microsoft SharePoint Foundation Web application has been created.

If this is the first time that you have used this application pool with a SharePoint Web application, you must wait until the Internet Information Services (IIS) Web site has been created on all servers. By default, no new SharePoint site collections are created with the Web application. If you have just created a Forms Based Authentication (FBA) Web application, then before creating a new site collection, you will need to perform some additional configuration steps.

Learn about how to configure a Web application for FBA.

Once you are finished, to create a new site collection, go to the Create Site Collection page.

OK

---

12. Enter a value for the **Title and Description** section of My Sites.
13. In the **Web Site Address** section, the URL you selected earlier for the My Site host should be displayed. Leave the drop-down menu for the path selected as / (since a root site collection is necessary for self-service site provisioning, which we'll do later).

14. In the **Template Selection** section, select the **Enterprise** tab and then select **My Site Host**:

15. In the **Primary Site Collection Administrator** section, specify a name for the owner of the site collection in `domain\username` format.

16. Click **OK** to begin provisioning.

17. Take note of the path where the site collection was created and click **OK**:

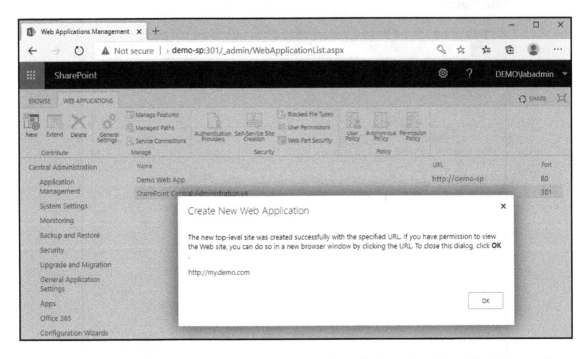

18. Select the new My Sites web application from the list of web applications so that it is highlighted.

19. From the ribbon, in the **Manage** group, select **Managed Paths**.

20. In the **Add a New Path** section, type in the name of the path you want to append to the My Sites URL namespace. For example, if you want a user's My Site address to be
   `http://my.domain.com/`*personal*`/<username>`, enter `personal` in the textbox. Click **Add Path** when you're finished, and then click **OK**:

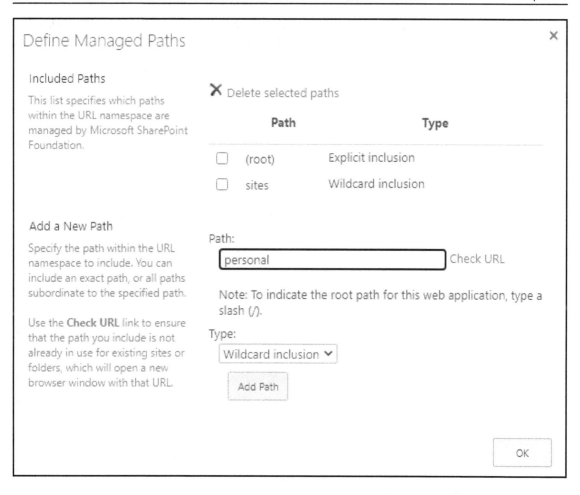

21. From the ribbon, in the **Security** group, select **Self-Service Site Creation**:

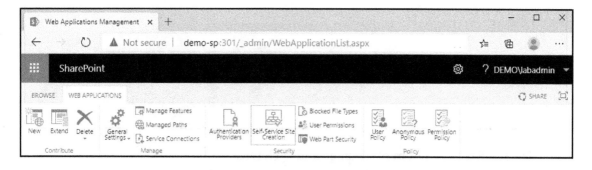

22. In the **Site Collection** section, select the **On** radio button. Leave the **Hide the Create site command** radio button selected:

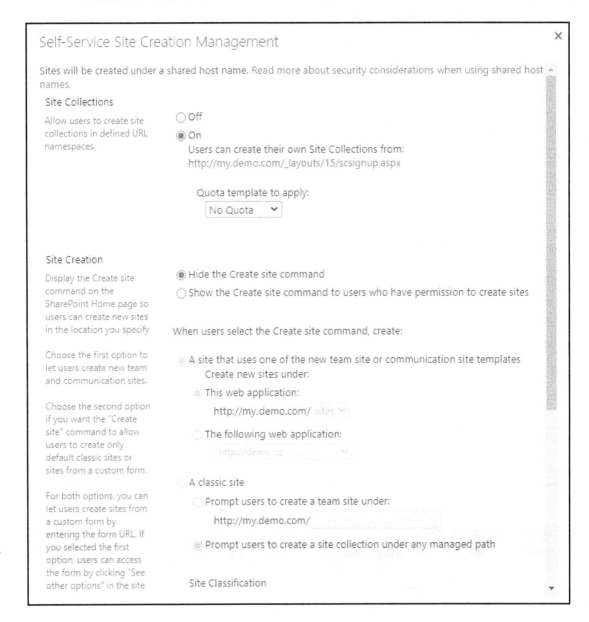

23. Scroll to the bottom of the page and click **OK**.
24. From the ribbon, in the **Policy** group, select **Permissions Policy**:

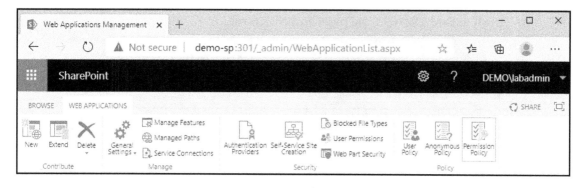

25. Click **Add Permission Policy Level**.

26. In the **Name and Description** section, add a name and, optionally, a description.

27. In the **Permissions** section, scroll to the **Site Permissions** subsection and select the checkbox in the **Grant** column for **Create Subsites - Create subsites such as team sites, Meeting Workspace sites, and Document Workspace sites**:

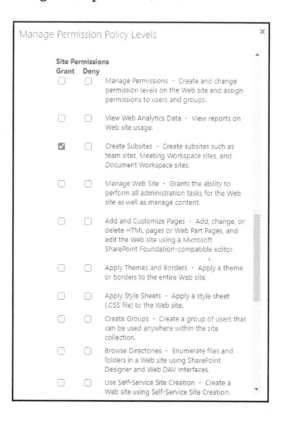

28. Scroll to the bottom of the page and click **Save**.
29. On the **Manage Permissions Policy Levels** dialog, click **OK**.
30. From the ribbon, in the **Policy** group, select **User Policy**.
31. Click **Add Users**:

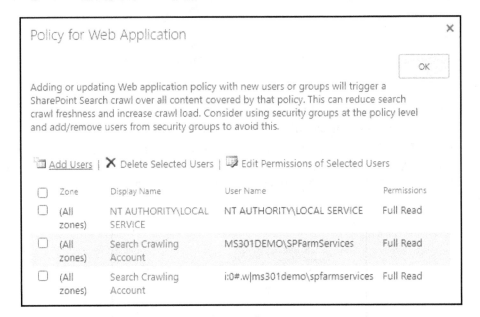

32. Under **Zones**, select **(All zones)** and click **Next**.
33. In the **Choose Users** section, add the users who you want to enable self-service My Sites provisioning for. For most organizations, this will be **All Users**. To select all users, click the **Browse** icon (which looks like an address book under the **Users** text area box):

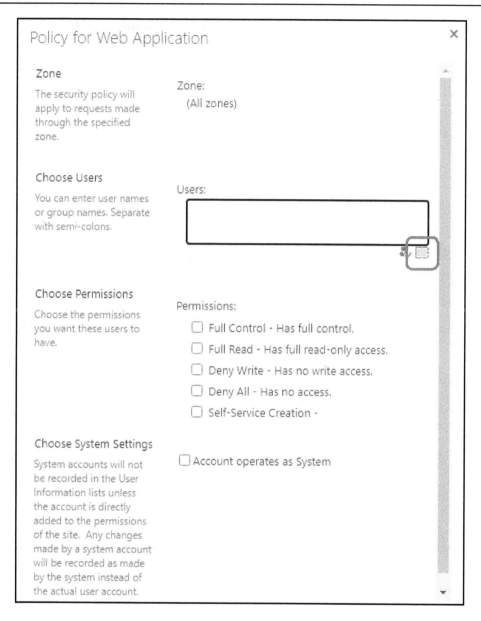

34. Select **All Users**, select **Everyone**, and then click the **Add ->** button. Click **OK** when you're finished:

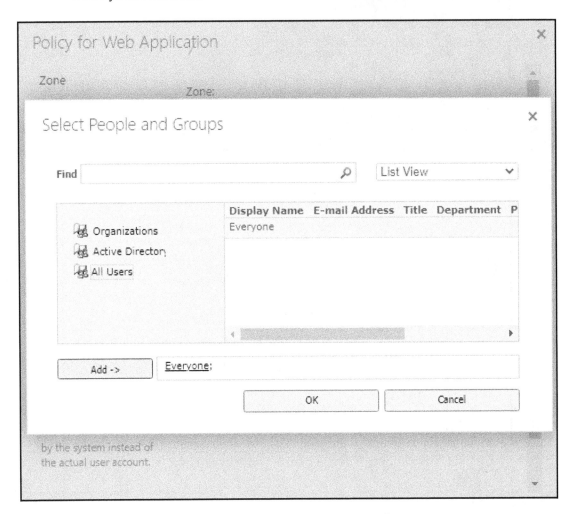

35. In the **Choose Permissions** section, select the self-service site creation policy you created previously, and then click **Finish**:

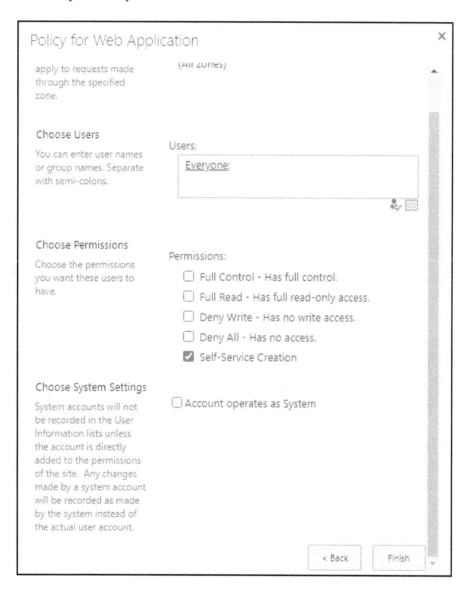

36. Note that the policy has been added. Click **OK**:

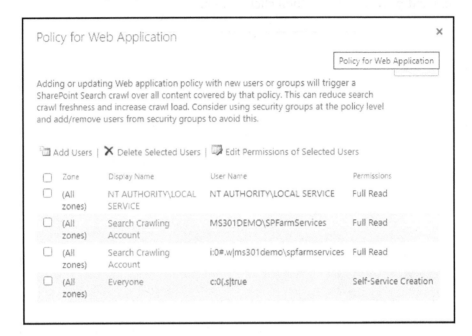

At this point, the My Sites host and site collection have been created and permissions have been assigned.

To finish this configuration, we'll need to update the User Profile service application.

Updating the User Profile service application

Once the My Site host site collection has been configured, you'll need to update the My Site settings in the User Profile service application to complete the configuration. These settings are necessary to ensure that the User Profile service application is aware of the correct My Site host collection settings.

To update the User Profile service application, follow these steps:

1. Launch **Central Administration**.
2. Click **Application Management**, and then select **Manage service applications**.
3. Select the User Profile service that you created previously so that it is highlighted.
4. From the ribbon, in the **Operations** section, choose **Manage**.
5. Under **My Sites Settings**, click **Setup My Sites**:

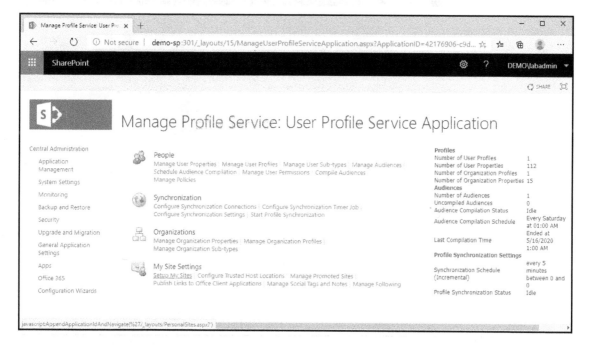

6. If you have a Search Center configured, you can enter the URL for it under the **Preferred Search Center** section. You'll learn more about configuring Search in Chapter 8, *Manage Search*.

7. In the **My Site Host** section, enter the URL of the My Site host site collection (*step 17* in the previous section).

8. In the **Personal Site Location** section, enter the managed path that you created (*step 20* in the previous section). Ensure that you exclude both leading and trailing slash characters:

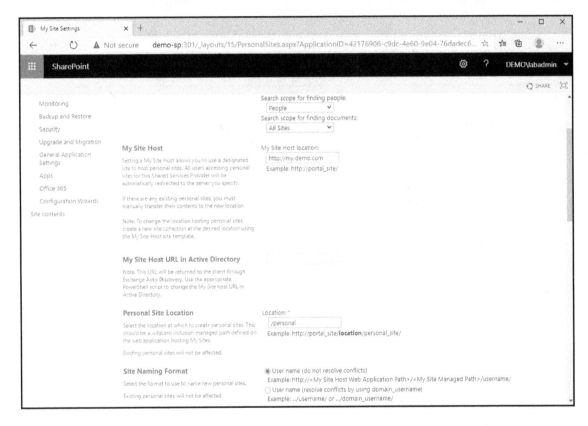

9. Under **Site Naming Format**, choose an option for generating the users' **My Site** path.

10. If you are configuring SharePoint Server 2016, you can set up some **Language Options** to allow the user to set the default language for their My Site. SharePoint Server 2019 defaults to the installation language for SharePoint.

11. In the **Read Permission Level** section, you may need to select names and click the **Resolve** icon to update them.

12. Under the **Newsfeed** section, you can enable or disable activities in the My Site newsfeed.

13. In the **Email Notifications** section, you can enter a reply address for system-generated emails.

14. In the **My Site Cleanup** section, leave the **Enable access delegation** checkbox selected. When a user is offboarded and their profile is deleted, this setting configures the user's manager (as listed in Active Directory) to receive access to their site. You can also specify an additional secondary owner for situations where a user's manager cannot be determined.

15. Under the **Privacy Settings** section, configure whether you want a user's My Site profile data to be public (followers and following, social tagging, birthdays, job title, blog posts, and other activities). If you make the sites public by default, users will still have the ability to update what things they want to be shared.

16. Click **OK**.

With that, the integration between My Sites and the User Profile service application is complete.

Next, we will discuss some of the monitoring capabilities of SharePoint Server.

Monitoring

As part of managing a SharePoint Server environment, you'll need to review data about its performance and statistics. Configuring the various logging options and reviewing the output periodically will help you understand how the environment is being used, as well as help you anticipate the needs for growth and detect potential issues.

In this section, we'll look at the SharePoint Health Analyzer, usage data, and diagnostic logging.

Monitoring farm Health Analyzer reports and resolving issues

The SharePoint Health Analyzer is a tool that is used to check for potential issues, including general configuration, performance, and usage. Using predefined rules that target security, performance, configuration, or availability, you can get a detailed report on the overall status of a farm.

When one or more servers, applications, or other entities exceeds thresholds defined by the rules, an alert may be displayed on the **Central Administration** dashboard, as shown in the following screenshot:

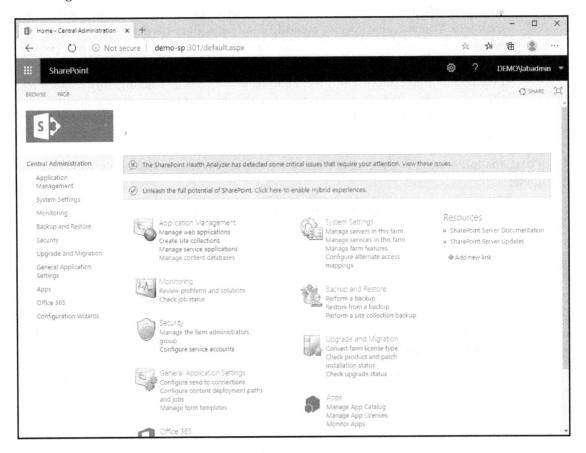

To view the conditions that are responsible for generating the error, click on the **View these issues** link. The health report will display the category, along with information on which services generated the report and which servers may be affected:

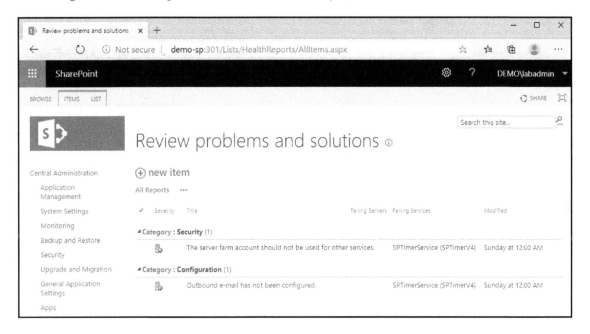

Each item in the list can be clicked on and will display additional information regarding the rule that generated the error, as well as potential remedies, as shown in the following screenshot:

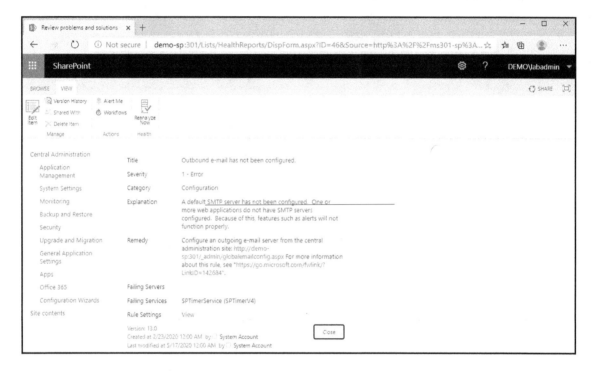

You can follow the steps listed under the **Remedy** section to resolve this issue. Once you have resolved the issue, you can click on the **Reanalyze Now** button in the **Health** group on the ribbon to test and verify the solution.

Configuring Health Analyzer rules

Depending on your business or organizational needs, you may find it necessary to adjust, enable, or disable particular Health Analyzer rules. To update the parameters for a particular rule, follow these steps:

1. Launch **Central Administration**.
2. Select **Monitoring**, and then select **Review rule definitions** under **Health Analyzer**:

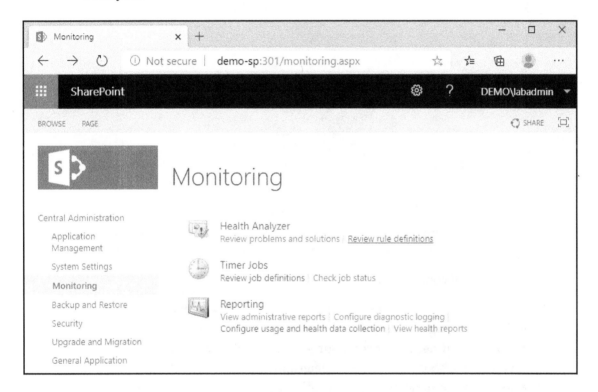

3. Select a rule to modify and click on it:

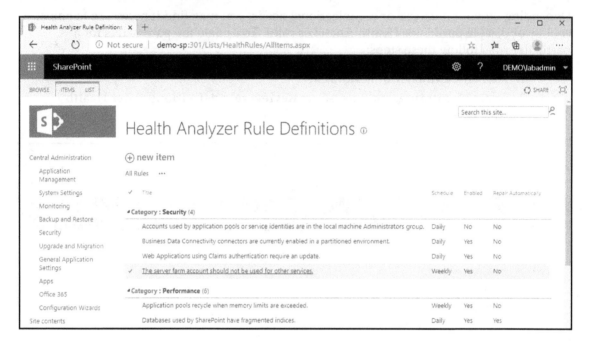

4. Click **Edit item** on the ribbon to customize the rule. Click **Save** when complete, or click **Cancel** to discard the changes you've made to the rule.

Each rule has several configurable fields:

- **Title**: The title is the description displayed in the health analyzer.
- **Scope**: Manages what server or servers to run the rule against.
- **Schedule**: Rules can be run hourly, daily, weekly, monthly, or on-demand.
- **Enabled**: Choose **Yes** or **No** to determine whether the rule runs.
- **Repair Automatically**: Some rules have built-in tasks that they can run to attempt to self-heal the problem. You can enable or disable this functionality. If a rule does not have any repair actions, this setting is ignored.
- **Version**: You can change the version number of the rule to indicate changes you've made to any of the configurable options.

While the preceding properties can be updated, the actual rules *cannot* be modified.

Configuring the Health Analyzer timer job

SharePoint Health Analyzer runs on a schedule that's controlled by a timer job. You can manage the schedule by following these steps:

1. Launch **Central Administration**.
2. Select **Monitoring**, and then, from the **Reporting** section, select **Configure usage and health data collection**.
3. Under the **Usage Data Selection** section, select **Enable usage data collection**:

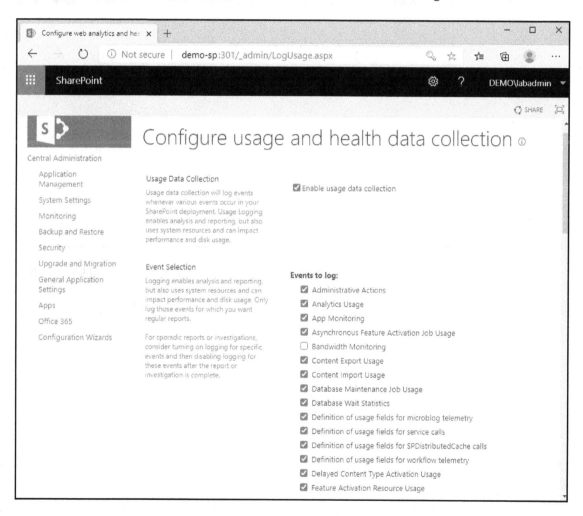

4. Scroll down to the **Health Data Collection** section and right-click the **Health Logging Schedule** link to the timer job definitions page. This will appear in a new tab:

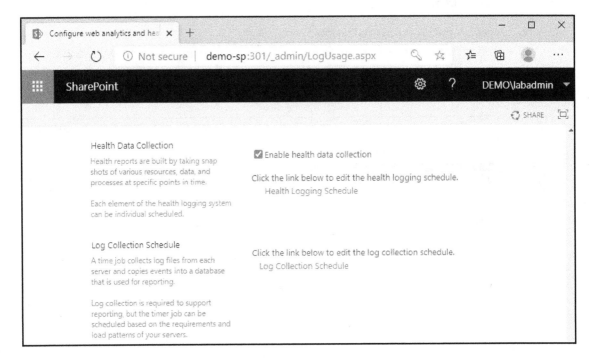

5. Scroll down to the **Health Analysis** jobs, and then click an individual job to edit its schedule:

6. Once you've updated the schedule of a particular job, click **OK** and close this tab.

7. Go back to the **Configure usage and health data collection** page and click **OK** to save any changes.

Monitoring storage usage for SharePoint

Usage reports are available for SharePoint sites and site collections. These usage reports show you what parts of a site are taking up space, both in real numbers and as a percentage of the site.

To access a storage metrics report for a site or site collection, follow these steps:

1. Navigate to a site.

2. Click the **Gear** icon, and then select **Site Settings**:

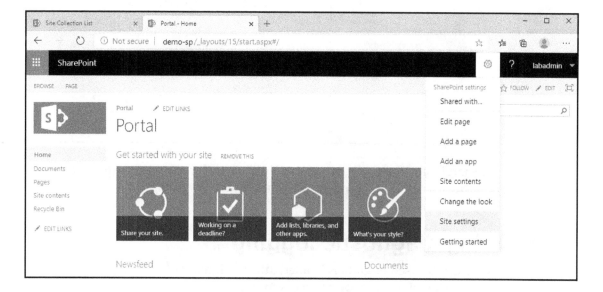

3. In the **Site Administration** section, select **Storage Metrics**.
4. Review the storage metrics data:

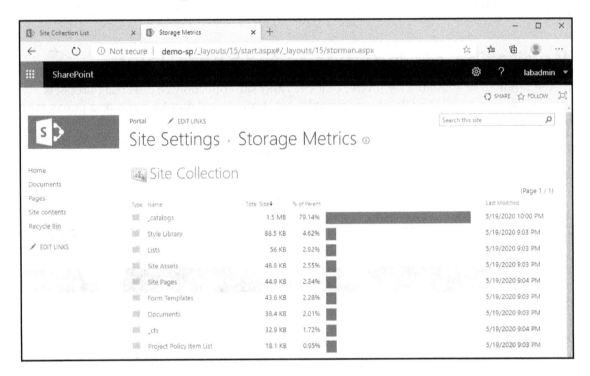

You can use the SharePoint storage reports in conjunction with reports from the SQL database infrastructure to estimate growth needs for capacity planning.

Configuring diagnostic logging

Diagnostic logging can be used to gather detailed information about the configuration, performance, and metrics of a SharePoint Server environment.

When configuring diagnostic logging, you can set the threshold for the detail level for data written to the server event and trace logs. The following tables describe the levels of logging available:

Event Log Detail Levels

Level	Definition
None	No logging is captured.
Critical	This is a serious error, resulting in a major failure.
Error	This is an urgent condition that requires investigation.
Warning	A potential problem or issue might require attention. Warnings should be tracked for any patterns.
Information	No action is required.
Verbose	This is a detailed information level for messages.

Trace Log Detail Levels

Level	Definition
None	No trace logs are captured.
Unexpected	This includes messages about events that cause solutions to stop processing. With this setting, the log will include events tagged at the Unexpected, Exception, Assert, and Critical levels.
Monitorable	This includes messages about all unrecoverable events that limit functionality but do not stop the application. With this setting, the log also includes events that the Unexpected setting records.
High	This includes messages about events that are unexpected but that do not stop a solution from being processed. This setting also includes all events that the Monitorable setting records.
Medium	This level records all high-level information about system operations. This level also includes all the events that the High setting records. It is the most common setting to use when you're beginning troubleshooting operations.
Verbose	When set to this level, the log includes most actions about operations in the system. This level is typically used only for debugging in a development environment. With this setting, the log will also include all events that the Medium setting records.
VerboseEx	This level can only be configured through the `Set-SPLogLevel` PowerShell cmdlet and includes very low-level diagnostic data. This level should only be used in a development environment. This log level includes everything from the Verbose setting as well.

To configure diagnostic logging, follow these steps:

1. Launch **Central Administration**.
2. Click **Monitoring**, and then, from the **Reporting** section, select **Configure diagnostic logging**.

3. Expand the list of categories and select one or more to enable logging. Scroll down and select the threshold of what log level to report for both the event and trace logs:

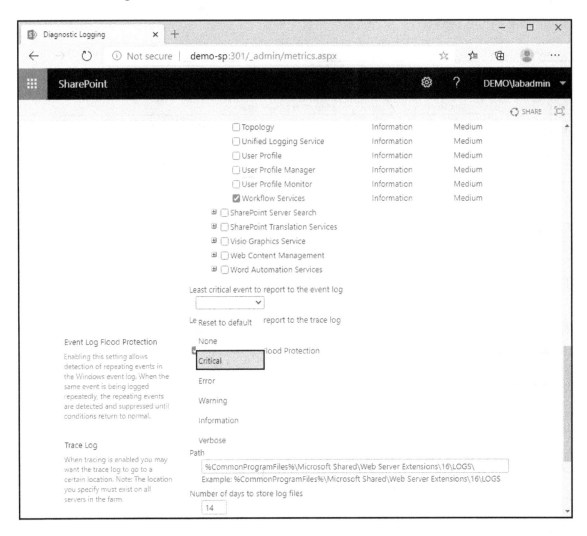

4. Under **Event Log Flood Protection**, select the **Enable Event Log Flood Protection** checkbox to limit the logging of duplicate events that appear.

5. Under **Trace Log**, enter a path where you wish to store the trace logs, a length of time to store the logs for, and a limit for consuming disk storage. Microsoft recommends that you limit the amount of trace log data you retain to prevent it from filling up the disk.

6. Click **OK** to complete the configuration.

Once logging has been configured, you can begin monitoring the event log. The log file path will store the trace logs, while the event log will store application events that are raised.

Next, we'll look at how to properly deploy updates to the SharePoint environment.

Updating SharePoint Server and validating the installation

At some point, every SharePoint Server farm needs to have maintenance tasks performed on it, such as updating with a newer service pack or cumulative update. These updates typically provide security enhancements or new features and capabilities.

Updating SharePoint Server farms can require multiple steps, including two separate types of update files:

- Public updates, which are language-independent updates
- A language-dependent patch, which contains language-specific updates

While the public updates are typically released monthly, the language-dependent patch may not be. When performing a public update, it's necessary to rerun the most recent language update patch to ensure that all the components continue to function correctly.

When software updates are released, you'll need to distribute the binaries (both the public update and the most recent language patch) to each server in the farm. There is not a specified server order – SharePoint updates are backward compatible to allow updated servers in the farm to coexist with servers that have yet to be updated. It is not recommended to run in such a configuration for extended periods of time, though.

The steps for performing updates on farms with and without high availability are different. Regardless of which farm topologies have been deployed, it's recommended to validate the updates and the process against a test environment before deploying to production.

While there are several different ways to approach the update cycle, including both complete installation/update cycles with downtime, staged update processes that reduce downtime (but don't eliminate it), or zero-downtime approaches with highly available farms, we'll focus on the most direct process, called the *in-place method without backward compatibility*.

For complex SharePoint environments or environments with minimal outage window limits, however, there are several additional considerations to enable updating with a very limited outage and handling a specific order of events. These processes are quite complex and require a detailed understanding of your SharePoint topology, as well as any load balancing technologies in use. SharePoint updating should typically be managed by SharePoint administrators due to the order and steps required and not treated simply with the "update and reboot" methodology. For more information on *zero-downtime patching*, see https://docs.microsoft.com/en-us/sharepoint/upgrade-and-update/sharepoint-server-2016-zero-downtime-patching-steps.

For our reference environment, we'll use the **Small High-Availability MinRole** server farm from Chapter 2, *Planning a SharePoint Farm*, as depicted in the following diagram:

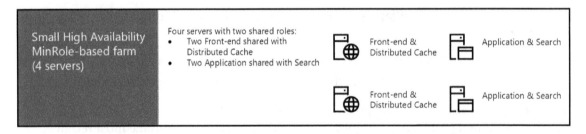

To perform the update, follow these steps:

1. Instruct users that the farm will be unavailable for the duration of the update.
2. Remove all **Front-end and Distributed Cache** servers from any load-balanced configuration.
3. On a server that runs Search components (in this case, one of the **Application and Search** servers), launch the SharePoint Management Shell and run the following command to pause **Search**:

```
$ssa=Get-SPEnterpriseSearchServiceApplication
Suspend-SPEnterpriseSearchServiceApplication -Identity $ssa
```

4. On each server that hosts one or more Search components, stop the following services (in this order):

- SPTimerV4
- OSearch16
- SPSearchHostController

5. Run the update executable on the **Application and Search** server hosting **Central Administration**.
6. Run the update executable on the remaining **Application and Search** servers.
7. On each server that hosts one or more Search components, start the following services (in this order):
   - SPSearchHostController
   - OSearch16
   - SPTimervV4

8. Verify that all search components are active by using the following command inside the SharePoint Management Shell:

```
Get-SPEnterpriseSearchStatus -SearchApplication $ssa | where
{$_.State -ne "Active"} | fl
```

9. Resume the Search service by running the following command inside the SharePoint Management Shell:

```
Resume-SPEnterpriseSearchServiceApplication -Identity $ssa
```

10. Log in to the first **Front-end and Distributed Cache** server.
11. Run the update executable on the first **Front-end and Distributed Cache** server.
12. Run the update executable on the remaining **Front-end and Distributed Cache** servers.
13. Run the SharePoint Products Configuration Wizard on the server hosting Central Administration. This will upgrade the configuration database and the content databases.
14. Run the SharePoint Products Configuration Wizard on the remaining **Application and Search** servers.

15. Run the SharePoint Products Configuration Wizard on the first **Front-end and Distributed Cache** server.
16. Run the SharePoint Products Configuration Wizard on the remaining **Front-end and Distributed Cache** servers.
17. Re-add the **Front-end and Distributed Cache** servers to any load balancer configuration.

With this, the farm has been updated.

You can further validate the success of the installation by reviewing the event logs. If errors occurred that are preventing updates, you can resolve the underlying conditions and resume or rerun the update as necessary.

For more complex update scenarios, including zero-downtime methods, see `https://docs.microsoft.com/en-us/sharepoint/upgrade-and-update/install-a-software-update`.

We'll expand the *updated* methodology to *upgrades* in the next section.

Selecting an upgrade path

If your SharePoint Server environment is not running on the 2019 platform yet, you will need to plan for an upgrade. If you are running SharePoint Server 2016, you may be able to perform a *database-attach* upgrade from 2016.

If you are migrating from SharePoint Server, the process is similar, but you'll need to perform a two-step migration (from SharePoint Server 2013 to SharePoint Server 2016, and then again from SharePoint Server 2016 to SharePoint Server 2019).

The following diagram depicts the high-level steps necessary to upgrade from SharePoint Server 2013 or 2016 to SharePoint Server 2019:

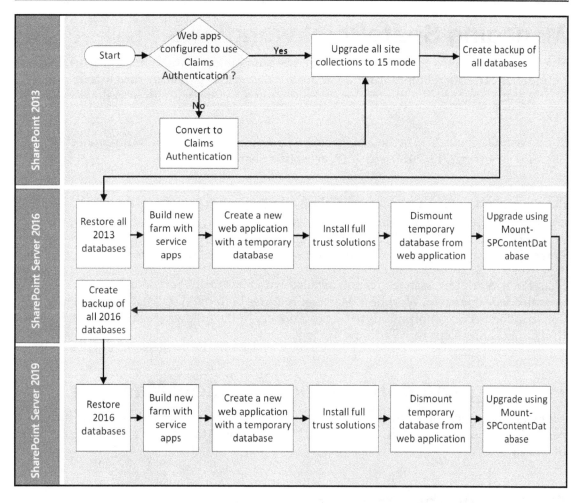

Depending on the size of the SharePoint farm and its activity level, organizations may also decide to opt for configuring SharePoint Hybrid environments and provisioning new workloads to SharePoint Online or migrating them altogether. You'll learn more about SharePoint Hybrid in `Chapter 10`, *Overview of SharePoint Hybrid*, while you learn more about migration options from `Chapter 16`, *Overview of the Migration Process*, onward.

Managing SharePoint workflows

SharePoint workflows are automation processes built on top of the Windows Workflow Foundation. The SharePoint Workflow Manager Platform is primarily developed with SharePoint Designer 2013, though Visual Studio can also interact with workflows to some degree.

SharePoint Designer 2013 is the last version of the designer platform. It can connect to SharePoint Server 2013, 2016, and 2019, as well as SharePoint Online.

Microsoft recommends utilizing the Power Automate platform for future workflow development. The Power Automate platform can connect to SharePoint Server environments using a data gateway. You'll learn more about data gateways in Chapter 14, *Implementing a Data Gateway.* You'll learn more about creating a workflow that uses the Power Platform in Chapter 15, *Using Power Automate with a Data Gateway.*

SharePoint Workflow Manager is not installed with SharePoint Server and must be downloaded separately. Workflow Manager requires both the App Management and Subscription Service applications to be provisioned. In the following sections, we'll look at the configuration steps for Workflow Manager.

Planning and configuring Workflow Manager

Workflow Manager can be installed on SharePoint farm servers, though it doesn't have to be. In this section, we will focus on configuring Workflow Manager on a server that is part of a farm.

Configuring the prerequisites

Installing Workflow Manager requires a number of prerequisites, such as creating service accounts and security controls. To configure these prerequisites, follow these steps:

1. On an Active Directory Domain Controller or computer with Remote Server Administration Tools, create an account for the workflow setup, such as **WorkflowSetup**. Take note of the password for later use.
2. Create an account for the workflow setup, such as **WorkflowSvc**. Take note of the account name and password for later use.

3. Create a security group for the workflow administrators, such as **Workflow Admins**.

4. Next, you'll need to add the SharePoint Server service accounts and workflow admin account and set up accounts for the Workflow Admins group you created in *step 3*.

5. Next, log into the server that is running SQL Server and launch **SQL Server Management Studio**.

6. Navigate to **Security** and then right-click and choose **New Login**:

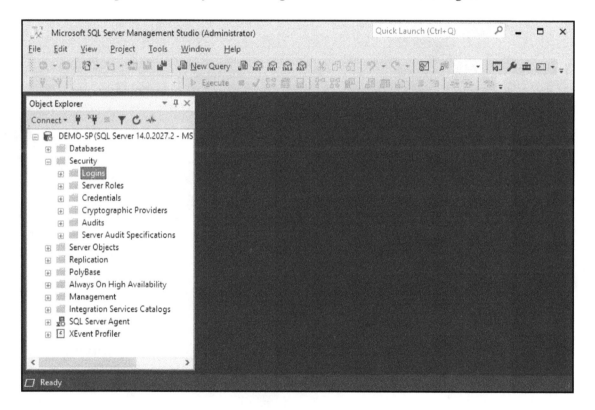

7. Search for and select the **WorkflowSetup** account you created in *step 1*:

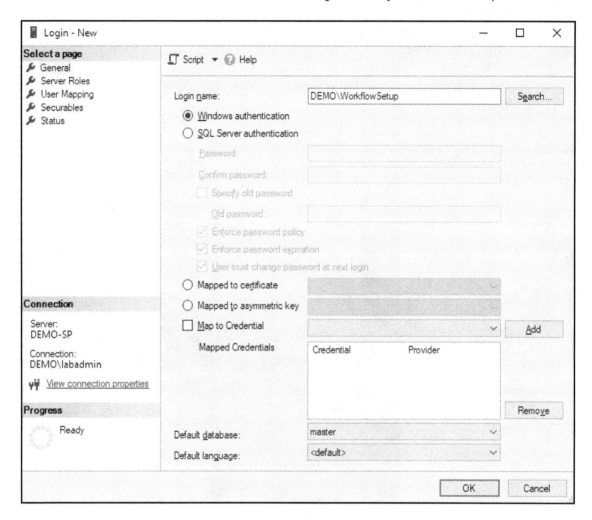

8. Select the **Server Roles** page, select the **sysadmin** checkbox, and then click **OK**:

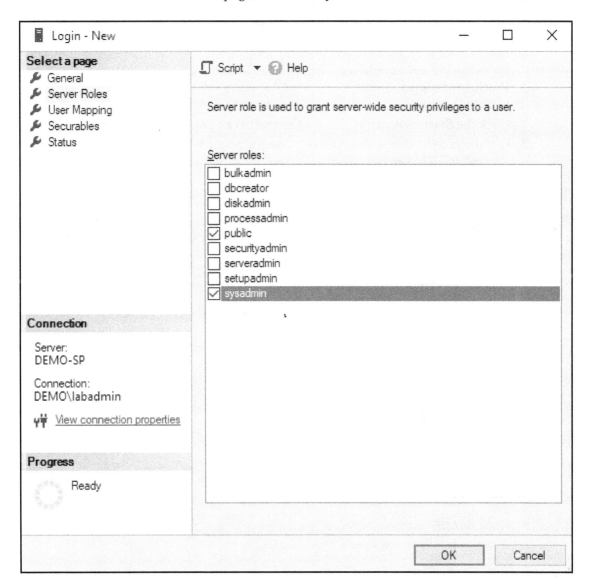

9. Finally, on the server where you are installing Workflow Manager, add the **WorkflowSetup** account to the local administrator's group.

Next, we'll begin the configuration of Workflow Manager.

Installing Workflow Manager

To install and configure Workflow Manager, follow these steps:

1. Log into the server where you're going to install Workflow Manager as the **WorkflowSetup** account (or a farm admin account, if you didn't create the **WorkflowSetup** account). Download the Microsoft Web Platform Installer from https://www.microsoft.com/web/Downloads/platform.aspx.
2. Click the **Start** menu and launch **Microsoft Web Platform Installer**.
3. In the search box, type **Workflow manager**. Select **Workflow Manager 1.0 Refresh (CU2)** from the list and click **Add**:

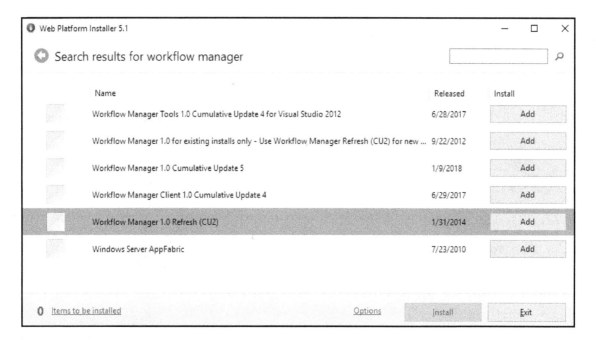

4. Click **Install**.
5. Accept the license terms and click **Install**.

6. Click **Continue** to launch the configuration wizard.

7. Click **Configure Workflow Manager with Default Settings (Recommended)**:

? – ×

WORKFLOW MANAGER CONFIGURATION WIZARD

Welcome

This wizard helps you configure a Workflow Manager farm. The wizard also configures the Service Bus farm required by the Workflow Manager farm.

⊙ **Configure Workflow Manager with Default Settings (Recommended)**
Apply default configuration settings and configure Workflow Manager on the machine. A Workflow Manager farm will be created and you can join other computers to the farm later.

⊙ **Configure Workflow Manager with Custom Settings**
Override default configuration settings to configure Workflow Manager on the machine. A Workflow Manager farm will be created and you can join other computers to the farm later.

⊙ **Join an Existing Workflow Manager Farm**
Choose this option to add this computer to existing Workflow Manager farm and Service Bus farm.

ⓘ Performing configuration operations simultaneously from multiple computers or configuration wizards is not supported.

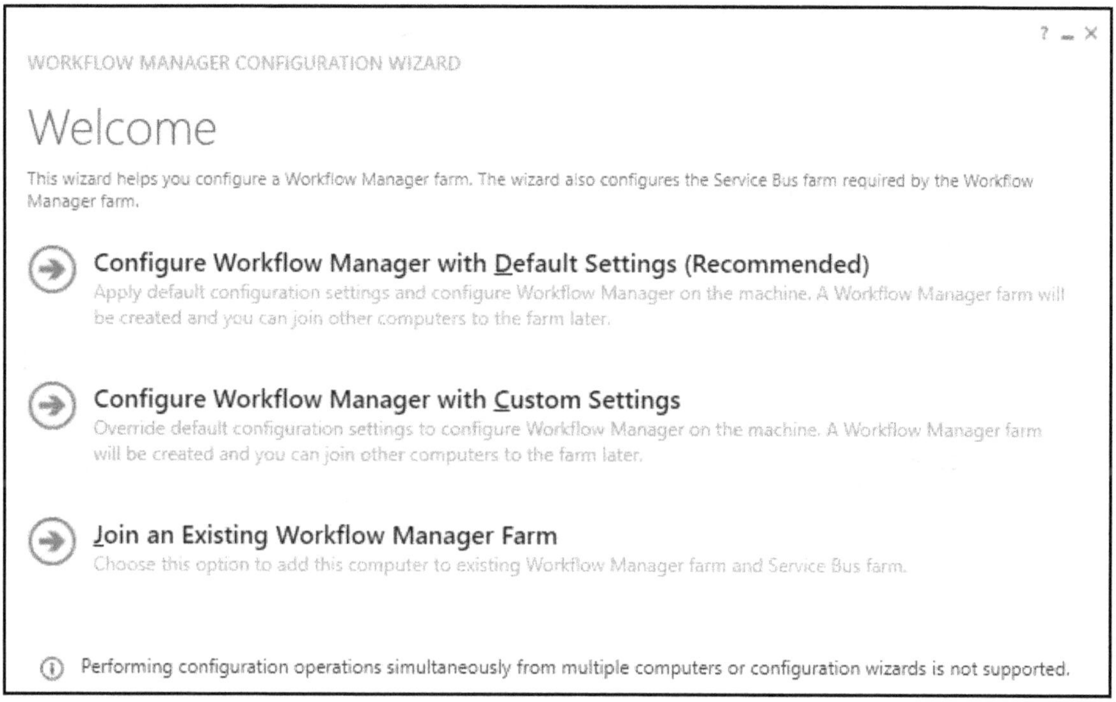

8. Enter the SQL Server name if necessary and click **Test Connection**.

9. Under **Configure Service Account**, update the user ID and password to that of the **Workflow Service** account you created previously. If you do not have an SSL certificate installed on the server, check the **Allow Workflow management over HTTP on this computer** box:

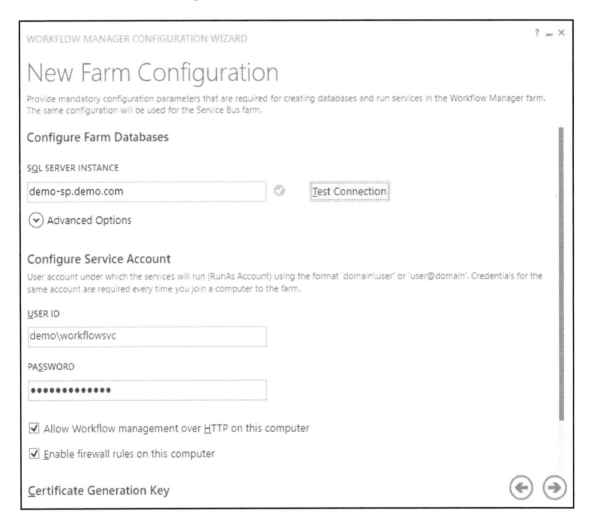

10. For **Certificate Generation Key**, enter a value that will be used when you join additional workflow servers.
11. Click the right arrow to proceed to the confirmation screen.

12. Review the configuration. Optionally, click **Copy** to copy the settings to the clipboard and save them:

13. Click the checkmark at the bottom of the page to begin the installation.
14. Review the installation log for any important messages and click the checkmark to close the installer.

Next, we'll install the Workflow Manager client on all servers that will participate with Workflow Manager.

Installing the Workflow Manager client

On each server in the SharePoint farm, download and install the Workflow Manager client. You'll need to add the **Workflow Setup** account you created previously to the local Administrators group on each server.

Once you have met these prerequisites, follow these steps to install the workflow client and join the farm:

1. Download the client from `https://go.microsoft.com/fwlink/p/?LinkID=268376`. Ensure that you just select the workflow client:

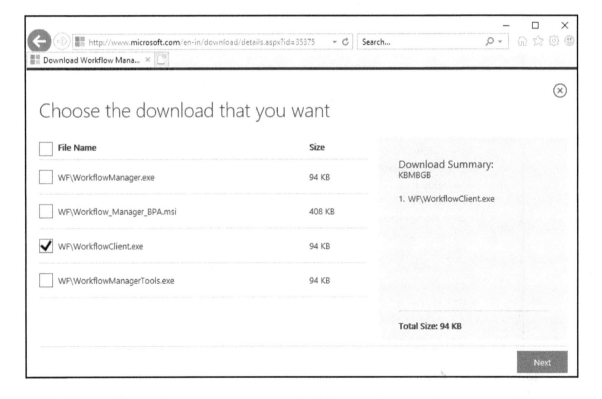

2. Launch the download and complete the Web Platform Installer. At the end of the installation process, click **Continue**.

3. Click **Join an Existing Workflow Manager Farm**:

4. Enter the SQL Server that Workflow Manager has been configured to use, click **Test Connection**, and then click the **Next** arrow:

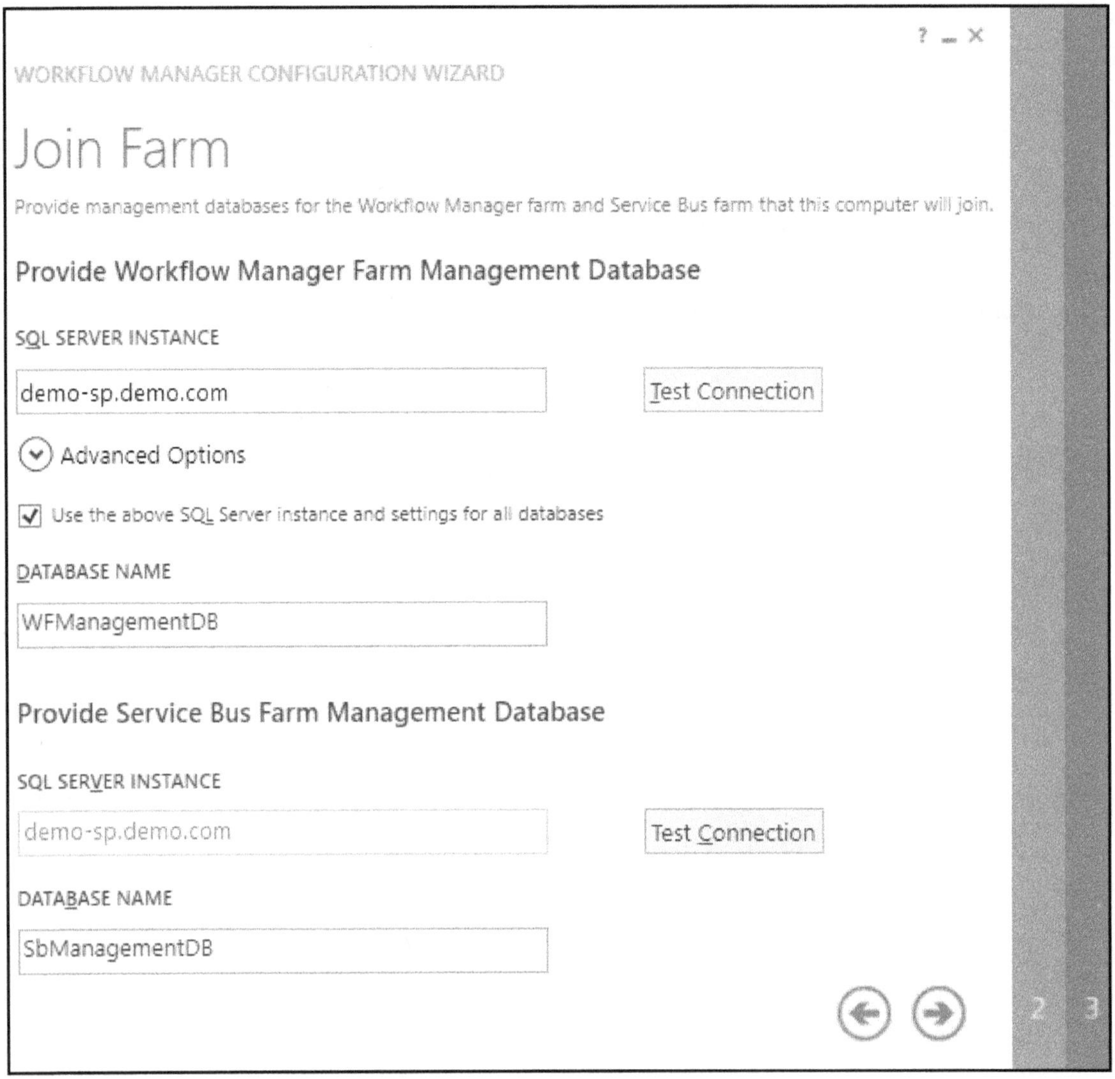

WORKFLOW MANAGER CONFIGURATION WIZARD

Join Farm

Provide management databases for the Workflow Manager farm and Service Bus farm that this computer will join.

Provide Workflow Manager Farm Management Database

SQL SERVER INSTANCE

| demo-sp.demo.com | Test Connection |

(⌄) Advanced Options

☑ Use the above SQL Server instance and settings for all databases

DATABASE NAME

| WFManagementDB |

Provide Service Bus Farm Management Database

SQL SERVER INSTANCE

| demo-sp.demo.com | Test Connection |

DATABASE NAME

| SbManagementDB |

(←) (→)   2 3

5. If the connection is successful, the service account name will automatically populate. Enter the password for the service account, as well as your **Certificate Generation Key**. Click the **Next** arrow when you're done:

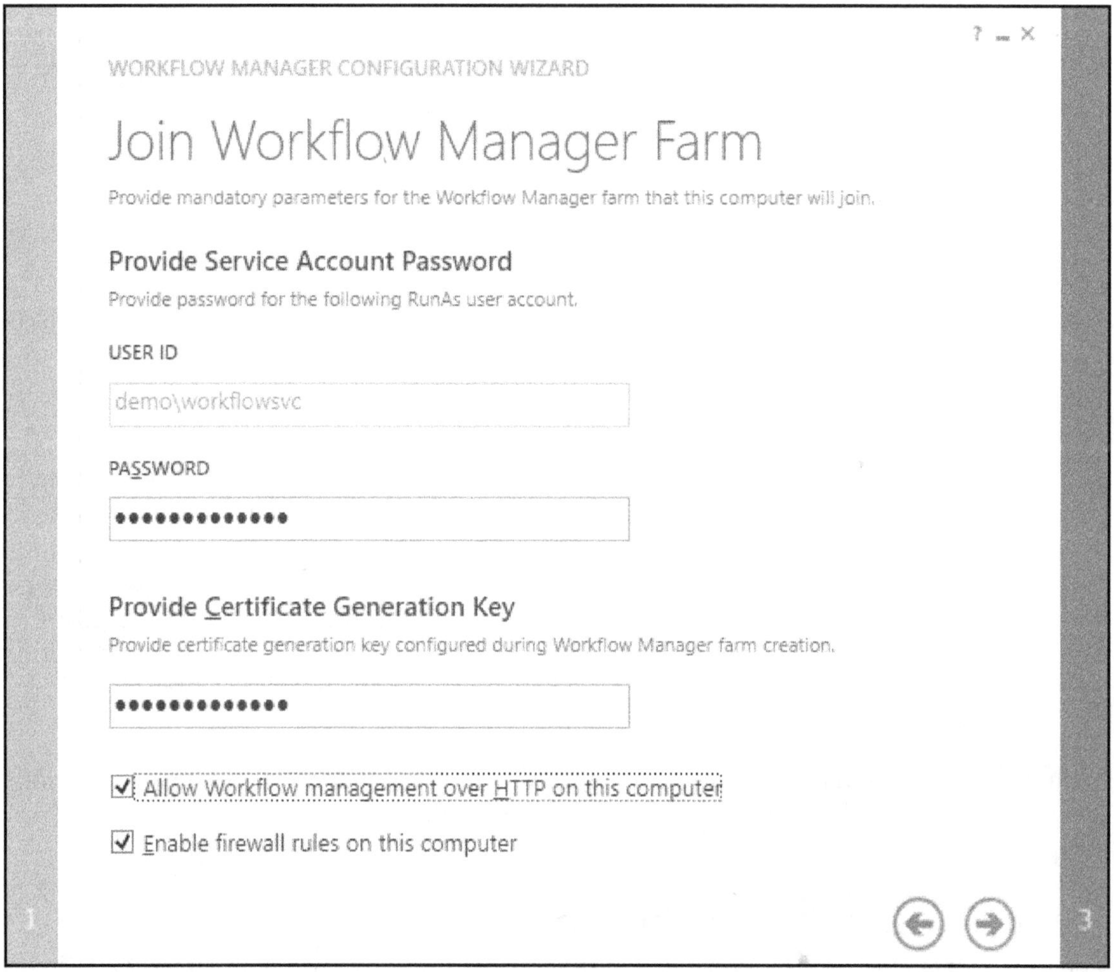

? _ X

WORKFLOW MANAGER CONFIGURATION WIZARD

Join Workflow Manager Farm

Provide mandatory parameters for the Workflow Manager farm that this computer will join.

Provide Service Account Password

Provide password for the following RunAs user account.

USER ID

demo\workflowsvc

PASSWORD

••••••••••••

Provide Certificate Generation Key

Provide certificate generation key configured during Workflow Manager farm creation.

••••••••••••

☑ Allow Workflow management over HTTP on this computer

☑ Enable firewall rules on this computer

(←) (→)

---

6. Select the option to use the same credentials on the Service Bus page. If you customized the original installation, you can enter the appropriate credentials for the Service Bus. Click the **Next** arrow when you're done:

7. Verify the settings and click the checkmark to begin the configuration.
8. Verify that the installation has completed successfully.

With that, the additional server has joined the Workflow Manager farm. In the next section, we'll connect the Workflow Manager farm to the SharePoint farm.

Connecting to SharePoint

The final step is making sure Workflow Manager is available for use with SharePoint. To do this, follow these steps on the server where Workflow Manager has been installed:

1. Launch **SharePoint Management Shell**.
2. Run the following command to get the Workflow farm and port:

```
$WorkflowUri = (Get-WFFarm).Endpoints[0]
```

3. Run the following command to register the farm with SharePoint, replacing `<site>` with a site collection where you want to use Workflow Manager:

```
Register-SPWorkflowService -SPSite https://<site> -WorkflowUri
$WorkflowUri
```

The Workflow farm should now be registered with SharePoint, allowing you to create and manage workflows.

Next, we'll look at some troubleshooting tasks for Workflow Manager.

Troubleshooting

Sometimes, when attempting to join a Workflow Manager farm, you may run into the issue depicted in the following screenshot:

This error indicates that the account you're attempting to configure Workflow Manager with is not a member of the Workflow Manager Service Bus users. You can check this by using the `Get-SBNameSpace` command from **SharePoint Management Shell**:

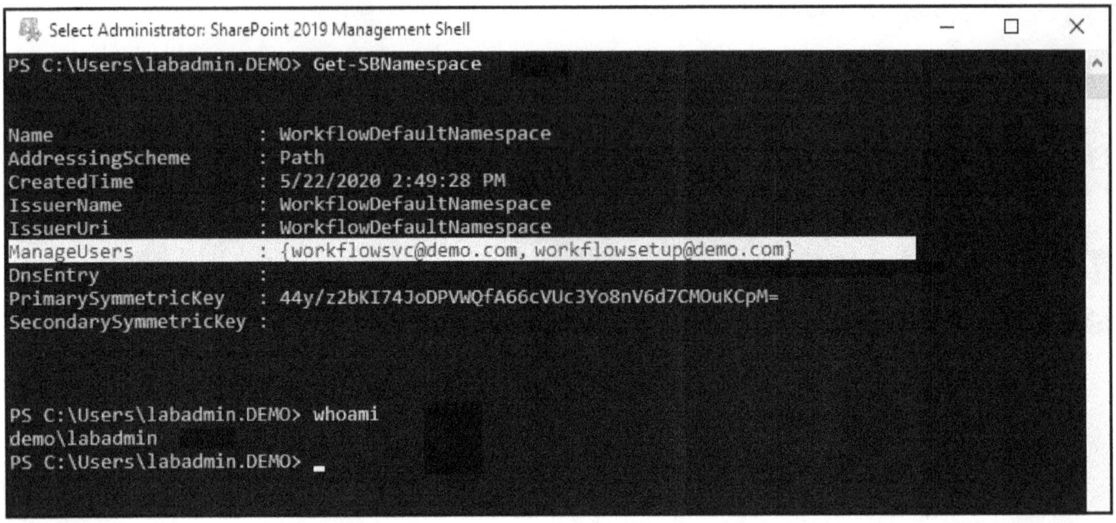

To resolve this issue, you can log off the server and log back in as one of the users listed under `ManageUsers`. Alternatively, you can update the users with the following command:

```
[array]$ManageUsers = (Get-SBNameSpace -Name
WorkflowDefaultNamespace).ManageUsers
$ManageUsers += "user@domain.com"
Set-SBNameSpace -Name WorkflowDefaultNamespace -ManageUsers $ManageUsers
```

The screen output should show that the `ManageUsers` parameter has been updated:

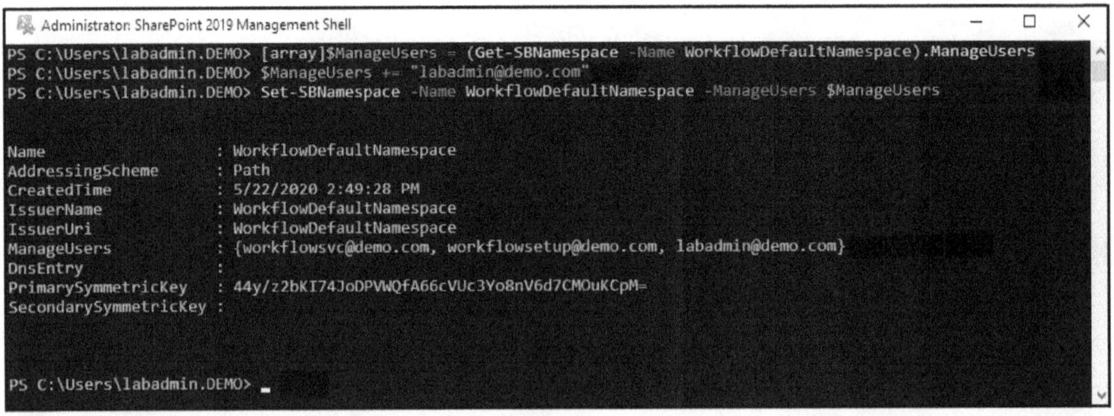

With that, you can rerun the **Join an Existing Workflow Manager Farm** task.

Next, we will learn about configuring a content type hub for SharePoint Server.

Configuring a content type hub

In SharePoint Server, a content type is a definition or a reusable collection of metadata, workflows, and other properties and settings for items in a SharePoint list or library. The benefit of a content type is that it allows you to manage settings for a class of information in a centralized way and reuse those parameters across a site.

A content type defines the following attributes of an item stored in SharePoint:

- Document templates (for document content types)
- Properties associated with the type
- Metadata associated with the type
- Information management policies associated with the type
- Workflows that can be started based on the type
- Custom features

A content type can be associated with a site or a library. When you associate a content type with a library or list, users can select that content type when creating new items. The content type can be used to guide users to specify certain properties of a document in a templated fashion or instantiate a workflow based on the item type.

Content types are local to the site where they are created. This is where a content type hub becomes important: it is a special site collection that's used to store content types and site columns. Content types can be used across site collections to drive uniformity and standardization.

In the following sections, we'll prepare the prerequisites and then create and configure a content type hub.

Configuring prerequisites

Before creating a content type hub, you need to meet the following prerequisites:

- You must have a managed metadata service
- You must have a site collection with the Team Site template

For information on configuring a managed metadata service, please follow the process outlined in Chapter 7, *Planning and Configuring Managed Metadata*.

For information on creating and configuring a site collection, please refer to Chapter 5, *Managing Site Collections*.

Setting up a content type hub

Once you have met the prerequisites for creating a content type hub, you can proceed with configuring the content type hub itself:

1. Launch a browser and navigate to the content type hub site collection.
2. Click the **Gear** icon, and then click **Site information**:

3. On the site information panel, select **View all site settings**.
4. Under **Site Collection Administration**, select **Site collection features**.
5. Locate the **Content Type Syndication Hub** feature and click **Activate**:

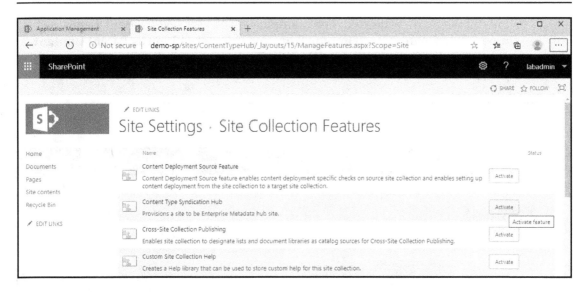

6. Click the browser's **Back** button to go back to the **Site Settings** page.
7. In the **Web Designer Galleries** section, select **Site content types**.
8. Click **Create** to create a new content type:

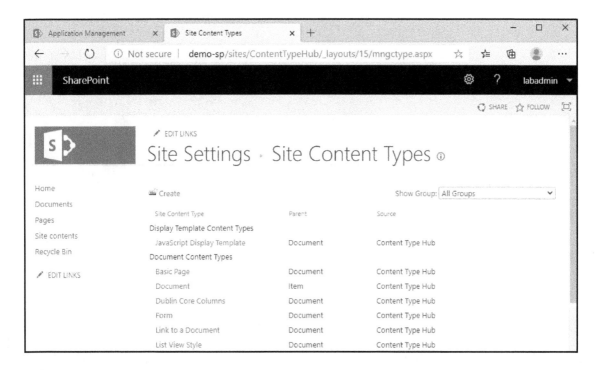

9. Fill out the fields as desired. For example, you can set **List Content Types** under **Select parent content type from** and then select **Item** as **Parent Content Type**. In this example, **Custom Content Type** has been selected as the type group:

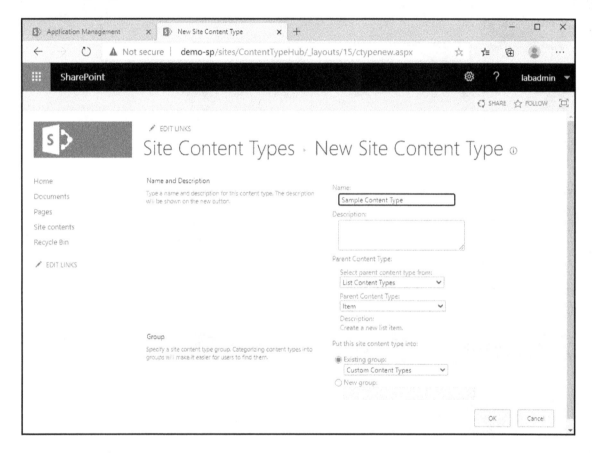

10. Click **OK** when you've finished.

11. The creation process will redirect you to the properties of the newly created content type. This is where you can edit these particular details:

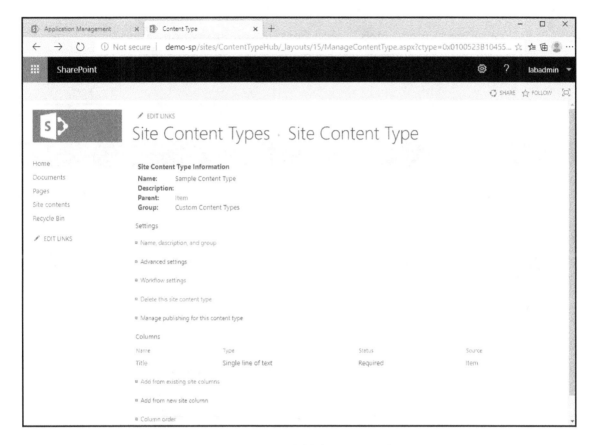

12. From this point on, you can click **Add from new site column** to create new fields for this content type, or **Add from existing site columns** to link existing columns to this content type.

Next, we'll configure a managed metadata service in order to consume this content type hub.

Configuring a Managed Metadata service

To configure a managed metadata service, follow these steps:

1. In a separate browser tab, launch **Central Administration**.
2. Navigate to **Application Management | Manage service applications**.
3. Highlight the **Managed Metadata Service** row and click **Properties**:

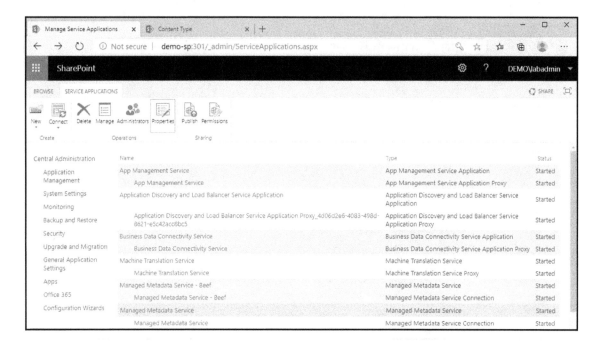

4. Locate the **Content Type Hub** section and enter the URL of the content type hub site collection:

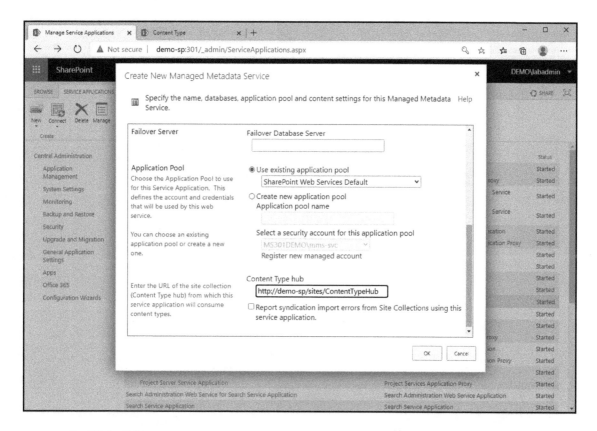

5. Click **OK**.
6. Click the **Managed Metadata service** sub-item and select **Properties**.

7. Select the **Consumes content types from the Content Type Hub Gallery at <site>** checkbox and click **OK**:

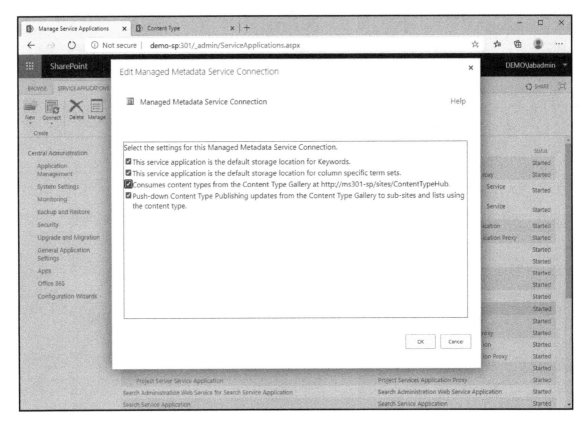

8. Select the browser tab that contains the content type you just created.

9. Select **Content Type Settings | Manage publishing for this content type**:

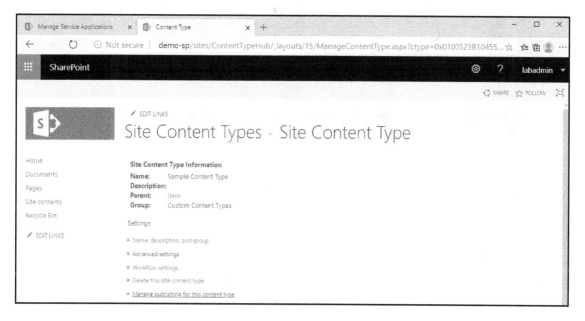

10. Select the **Publish** radio button and then click **OK**.

11. Select the browser tab where you had **Central Administration** open. Click **Monitoring | Review job definitions**.

12. Click on each of the content type hub jobs and then click **Run** to initiate the timer jobs to be published.

13. On a site where you will use the content type, navigate to **Site Settings** and click **Content Type Publishing**.

14. Under the **Refresh All Published Content Types** section, select the **Refresh all published content types on next update** checkbox and click **OK**. The sample content type that was created previously should show up as a **Subscribed Content Type**:

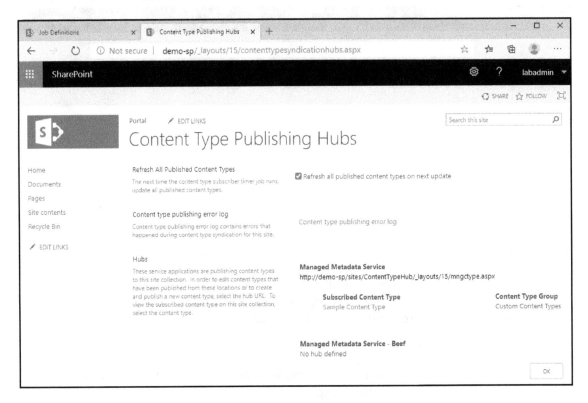

The content type will now be available for use on the site.

Troubleshooting performance issues

Since much of SharePoint Server's activity is storing or retrieving data from SQL Server, most performance troubleshooting and improvement tasks will rely on troubleshooting the underlying SQL environment.

Core recommendations for improving the performance of a SharePoint farm include the following:

- Ensuring an appropriate amount of frontend and distributed cache servers are deployed with adequate amounts of memory
- Validating the storage architecture for SQL is adequate for the number of read/write IOPS
- Validating all servers in the farm have adequate memory configurations using system performance monitoring for memory utilization and memory paging operations

For detailed recommendations on configuring SharePoint and SQL for best performance, see `Chapter 2`, *Planning a SharePoint Farm*.

Configuring SMTP authentication for a SharePoint farm

A feature that's new to SharePoint Server 2019 is SMTP authentication. In previous versions of SharePoint Server, servers in the farm could only send messages through an existing anonymous SMTP relay. With SMTP authentication, servers can log into a solution and send emails directly (such as through Exchange Online Protection or other systems).

SMTP authentication requires two configuration tasks to be carried out:

- Configuring an application credential key on each server in the farm
- Configuring outgoing emails for the farm

To configure SMTP authentication, follow these steps on each server in the farm:

1. Choose an application credential key to use for the farm. This key will be used to encrypt and decrypt the stored password. This key *must be the same on all servers*.
2. Launch **SharePoint Management Shell**.
3. Run the following PowerShell commands, replacing the `<ApplicationCredentialKeyValue>` value with the application credential key you chose in *step 1*:

```
$data = "<ApplicationCredentialKeyValue>"
$key = ConvertTo-SecureString -String $data -AsPlainText -Force
Set-SPApplicationCredentialKey -Password $key
```

Once you've done this, you can configure outgoing emails on the farm. Follow these steps to do so:

1. Launch **Central Administration**.
2. Select **System Settings** and then, under **E-Mail and Text Messages (SMS)**, select **Configure outgoing e-mail settings**:

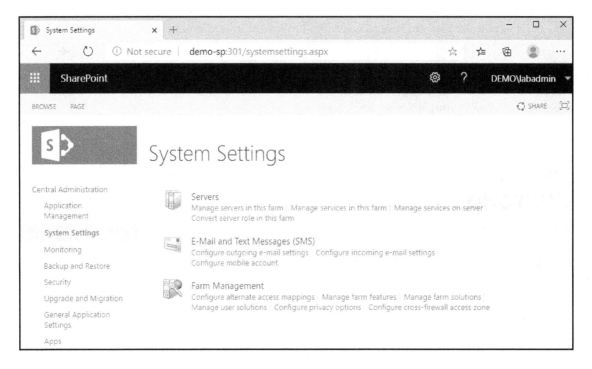

3. Under the **Mail Settings** section, enter the fully qualified server name or IP address for outbound mail in the **Outbound SMTP Server** box.
4. Enter the port the specified server uses. If no port is specified, SharePoint Server will use a default of port 25.
5. Enter the values for **From address** and **Reply-to address** that will be used for outbound messages.
6. Select an appropriate character set. The default for English-based systems is **65001 (Unicode UTF-8)**, but depending on the configured language for your system, you may need to choose another character set.
7. Under the **Mail Security** section, select the **Authenticated** radio button, and then enter the appropriate credentials.

8. If your outbound server requires it, select the **Yes** radio button to enable **TLS**:

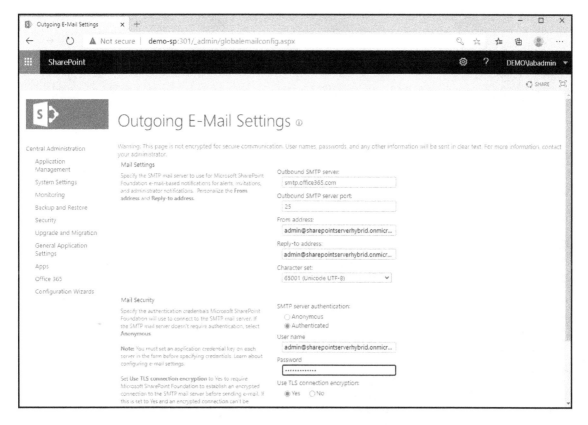

9. Click **OK** to save the configuration.

If you are specifying a **From address** that is different to the one you used under the **Mail security** section, you'll need to ensure that the selected credential has been granted *Send-As* permissions for the address specified in the **From address** section.

You can test your configuration in PowerShell by using a script similar to the following:

```
Add-PSSnapin Microsoft.SharePoint.Powershell
$Url = (Get-SPSite -Limit 1 -wa sil).Url
$Web = (Get-SPWeb $Url)
While ($To -eq $null) { $To = (Read-Host "Enter recipient email:") }
$Subject = "Test SharePoint Server Authenticated Submission"
$Body = "This is a test of the SharePoint Server 2019 Email with
Authenticated Submission"
$Send =
[Microsoft.SharePoint.Utilities.SPUtility]::SendEmail($Web,0,0,$To,$Subject
,$Body)
```

If successful, the specified recipient should receive an email whose subject is **Test SharePoint Server Authenticated Submission**.

Summary

In this chapter, you learned about a lot of concepts, including how to plan for and configure Distributed Cache, the User Profile service application, My Sites, Workflow Manager, web applications, and the content type hub. Each of these areas is fundamental to the performance and usability of a farm. Many of the configurations are only performed once, but they must be done correctly to ensure the farm has a good foundation.

One of the most important topics we covered was configuring My Sites and the User Profile service – these two services are critical for helping end users get the most value out of SharePoint Server. The SharePoint User Profile service can be used to aggregate data from many systems. It then uses that data to build a robust interface for users to interact with their personal data. The User Profile service and My Sites applications are interconnected, so when you make modifications to one, you have to be aware of implications on the other.

You also learned about some of the best practices around configuring host-named site collections, which is Microsoft's recommended architecture moving forward.

Finally, you learned about configuring a feature new to SharePoint Server 2019 – authenticated SMTP for delivering messages to users.

In the next chapter, you'll learn about implementing authentication mechanisms for SharePoint Server, which allows you to provide security and access control to sites and applications.

4
Implementing Authentication

Most organizations have certain data that is designated for "authorized users only."
Whether this is critical sales data, engineering and design samples, market research, human
resources information, or other privileged information, it's important to make sure that
only the correct users are allowed access.

This process of validating a user is called **authentication**. The authentication process takes
the credentials submitted by a user and checks them against some sort of user account
database. A user account database, in this context, is referred to as the **authentication
provider**, and it can be an active directory, an LDAP, Azure Active Directory, or some other
directory or database structure.

In this chapter, we will examine the various authentication aspects of SharePoint Server,
including the following:

- Selecting and implementing an authentication method
- Configuring web apps with multiple authentication methods
- Monitoring and maintaining authentication
- Troubleshooting authentication issues

After you've completed this chapter, you should be able to identify the methods used to
authenticate users and be familiar with the processes of configuring various authentication
mechanisms and troubleshooting authentication issues.

Selecting and implementing an authentication method

During the authentication process, a user asserts their identity. Once their identity is validated, the authentication provider's proof of the user's assertion is returned (usually as some sort of token). The result of this claims-based authentication system is a claims-based security token, issued by the SharePoint security token service. SharePoint Server utilizes this claims-based authentication process to provide access to resources.

When planning and configuring authentication methods, you'll need to determine which types you'll use for which web applications, which identity store will host the user identities, and the necessary supporting infrastructure for identity management.

SharePoint Server supports Windows, forms, and **Security Assertion Markup Language** (**SAML**)-based claims authentication. SharePoint also supports Kerberos-based authentication.

SharePoint also supports a mode known as *classic mode authentication*. Classic mode authentication is not compatible with server-to-server and app authentication and is, therefore, highly discouraged. Classic mode authentication cannot be configured through central administration and can only be configured through Windows PowerShell.

In the following sections, we'll discuss some of the authentication methods and terminology.

Authentication methods

Modern authentication design principles primarily focus on the various forms of claims-based authentication. Microsoft recommends using claims-based authentication for all web applications and zones in a SharePoint farm. The following diagram shows an overview of how authentication works in a SharePoint farm:

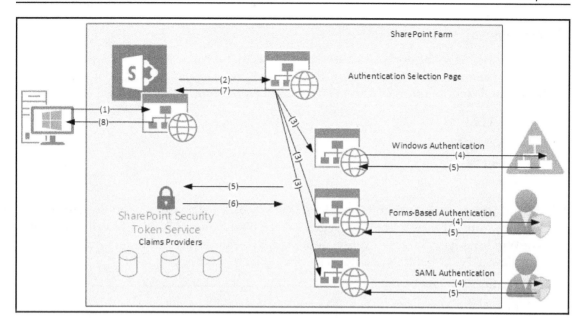

The steps are as follows:

1. The client requests a SharePoint resource.
2. The request is routed to the authentication components based on the settings for the zone.
3. The request is processed by the authentication components. If more than one authentication method is configured, the user selects an authentication method.
4. The user is authenticated by the identity provider.
5. If the authentication is successful, the SharePoint security token service generates a claims-based token for the user.
6. The claims-based token is sent back to the authentication components.
7. The authentication components redirect the request back to the resource with the claims-based token attached.
8. The request is completed and sent back to the user.

We'll now look briefly at the features of each of the other authentication types.

Windows

One of the advantages of Windows authentication is that it can use the existing **Active Directory Domain Services (AD DS)** authentication provider infrastructure in your environment. Windows authentication supports the following protocols:

- **NTLM**
- **Kerberos**
- **Digest**
- **Basic**

NTLM and Kerberos are both integrated into the Windows platform and allow users to access SharePoint services without providing additional credentials. NTLM is the simplest form of Windows-based authentication and typically requires no additional configuration. An overview of NTLM authentication is shown in the following diagram:

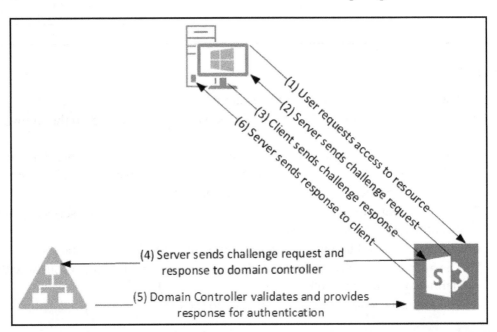

Kerberos is also integrated into the Windows platform but does require additional configuration, such as registering **Service Principal Names (SPNs)** before configuring a web application. Kerberos also requires the client computer to have connectivity to both a domain controller and the Kerberos key distribution center, reducing its usability for internet-facing scenarios. The following simplified diagram illustrates the Kerberos process:

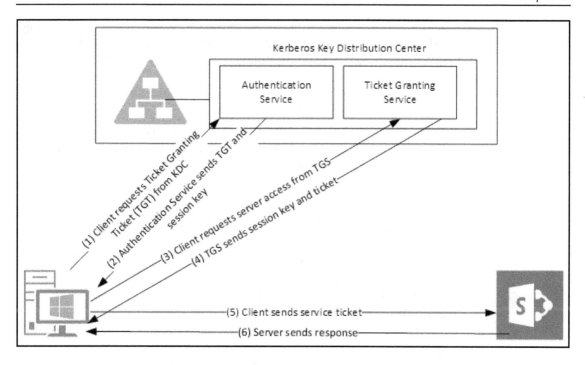

The following steps need to be taken to set up Kerberos:

1. Ensure that the name resolution for the web app to be configured exists. When configuring a web app, such as `portal.contoso.com`, you must be able to resolve that name in the DNS. If no record exists, create one that points to the server that will host the web app.

2. Configure HTTP SPNs for the web app URL and the application pool service account. From the Command Prompt, while logged in as a domain administrator, run the following commands:

```
SetSPN -S HTTP/<webapp> <DOMAIN\ServiceAccount>
SetSPN -S HTTP/<webapp.fqdn> <DOMAIN\ServiceAccount>
```

For example, if your service account is `SPWebApp` and your new web app is `portal.contoso.com`, the command will be as follows:

```
SetSPN -S HTTP/portal CONTOSO\SPWebApp
SetSPN -S HTTP/portal.contoso.com CONTOSO\SPWebApp
```

3. Trust the server for delegation (which is optional if only local users will be using local applications; if you are going to use report services or other things that reside outside of the SharePoint environment, you'll need to configure **Delegation**):
    - Launch **Active Directory Users and Computers** and locate the SharePoint server.
    - Right-click and go to **Properties** on the server object.
    - From the **Delegation** tab, select **Trust the computer for delegation to any service (Kerberos only)**.

4. Edit the authentication providers in the web application (as seen in Chapter 3, *Managing and Maintaining a SharePoint Farm*) and select **Enable Windows Authentication** and the **Integrated Windows Authentication** sub-checkbox. Select **Negotiate (Kerberos)** from the list.

5. Update or verify the configuration for the IIS web app:
    - Launch **IIS Manager**, locate the website for the web application, and select **Authentication** from **Features View**.
    - Select **Windows Authentication**, and then select **Providers**.
    - Verify that **Negotiate** and **NTLM** are listed (in that order). Click **Cancel**.
    - Click **Advanced Settings...** and under **Extended Protection**, ensure that the selection is set to **Off**, and then ensure that **Enable Kernel-mode authentication** is unchecked. Click **Cancel**.

The digest and basic authentication methods are older and less secure. Digest authentication sends user credentials as an MD5 digest to the web server; basic authentication sends credentials in cleartext and should be avoided unless traffic is secured with an SSL certificate.

Forms

Most users are familiar with forms-based credentials—they're used when the client computer doesn't have an integrated or secure relationship with the host computers. These are frequently used in internet scenarios. A forms-based authentication system frequently uses a membership directory or provider that is outside of AD DS.

In a forms-based authentication flow, the following process occurs:

1. The user requests a SharePoint resource.
2. SharePoint redirects the user to the forms-based authentication logon page.
3. The user's credentials are submitted to the SharePoint security token service.
4. The credentials are validated with the identity and membership provider. If successful, the appropriate claims are added to the user's token.
5. The token is issued to the user.
6. The request is sent to any other components in the authorization pipeline for the web app.
7. The response is sent back to the user.

This authentication flow is common for users that are members of an ASP.NET membership database or other database structures.

SAML

SAML-based configurations are also supported and require additional work to configure. You might use a SAML-based authentication framework if you're connecting with a partner organization that has its own identity provider.

SAML frameworks require and utilize several components:

- **Realm**: The URL or URI of a web application.
- **The SharePoint security token service**: The security token service is used to create and manage tokens for server-to-server authentication and claims for Windows, forms, and SAML authentications.
- **Identity Provider Security Token Service (IP-STS)**: A secure token service that issues SAML tokens on behalf of users in the associated directory.
- **Relying Party Security Token Service (RP-STS)**: Each web application is registered in the IP-STS server as an RP-STS entry.
- `SPTrustedIdentityTokenIssuer`: A special farm object that governs the communication to create and receive tokens from the IP-STS.

- **Token-signing certificate**: A certificate exported from IP-STS that is then used to create `SPTrustedIdentityTokenIssuer`.
- **Identity claim**: A SAML token claim that represents the unique identity of the user.
- **Other claims**: Additional claims that describe users, such as roles, groups, or other information.

SAML-based authentication takes the following process:

1. The user requests a SharePoint resource.
2. SharePoint redirects to the SAML authentication page.
3. Depending on the login provider, the request is redirected to the enterprise or federated STS login page.
4. The user provides credentials. If successful, the STS issues a SAML claims-based token.
5. A claims-based token is requested from the SharePoint STS, using the claims-based token from the external STS as authentication proof.
6. The SharePoint STS issues a claims-based token and any other additional claims if another claims provider is registered for the web application or zone.
7. The request is processed by any other authentication and authorization components configured in the pipeline.
8. The response is sent back to the user.

Let's talk about zones in the next section.

Zones

Zones are logical constructs that are used to gain access to sites in a web application. Each web application has a *default* zone but can be *extended* to support up to five zones. You can implement multiple authentication providers in a single zone; you can also use multiple zones with different authentication providers.

Microsoft generally recommends implementing multiple authentication methods on the default zone, resulting in a single URL for all users. If you choose to implement multiple zones, keep the following design constraints in mind:

- You can only deploy one instance of forms-based authentication on a zone.
- At least one zone must have NTLM as an authentication method for content crawling.

In the next section, we'll look at configuring authentication for internal and external users and assigning multiple authentication methods to a web application.

Configuring web apps with multiple authentication methods

As mentioned in the previous section, a web application can have multiple authentication providers. Additionally, authentication providers can get their user databases from a number of different places, including from Active Directory, a SQL database, or another federated environment.

In this example, we're going to create a new web application (although this process can also be used for an existing web application) and connect it to both an internal authentication provider (NTLM) an external identity provider that's been configured (the ASP.NET IIS authentication database).

To configure a web app to use multiple providers, follow these steps:

1. Launch **Central Administration** and navigate to **Application Management** | **Manage Web Applications**.
2. From the **Contribute** group on the ribbon, select **New**.
3. In this example, create a new IIS website for the web application. Specify a site name, port, and path. Leave anonymous access off (as well as SSL).

4. Since the goal is to configure multiple authentication providers, select the **Enable Windows Authentication** checkbox:

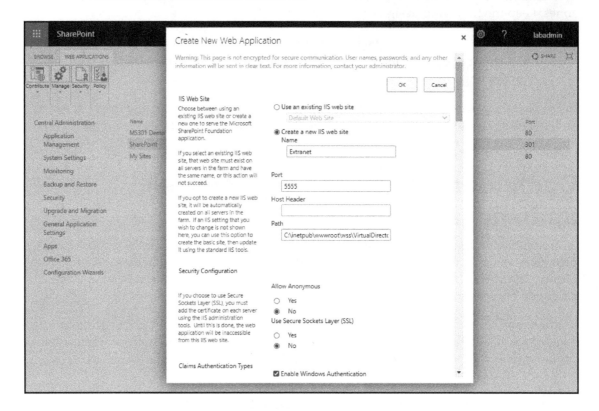

5. Select the **Enable Forms Based Authentication (FBA)** checkbox. Enter the **Forms-Based Authentication** membership and role providers of your application.

6. Specify a URL, application pool, and security settings (refer to `Chapter 3, Managing and Maintaining a SharePoint Farm`, for more details).

7. Specify the database settings.

8. When you have finished, click **OK**.

9. After the web application has been configured, click on **Create Site Collection**.

10. Under **Web Application**, select the newly created web application.

11. Enter a title and description.

12. Select a website address path, a template, and the administrators. Click **OK**.
13. After the site collection has been created, click the site link to open the site.
14. Click **OK** to close the window.

Once the web app has been provisioned, you can navigate to it. You should be prompted for an authentication mechanism:

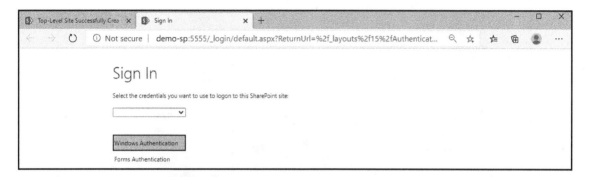

Selecting either **Forms Authentication** or **Windows Authentication** will direct you to the appropriate mechanism.   Next, we'll look at monitoring authentication.

Monitoring and maintaining authentication

If applications are unable to authenticate users, you can configure additional details for login capture. Following the process that we covered in Chapter 3, *Managing and Maintaining a SharePoint Farm*, you can specifically target authentication logging to help determine the source of errors.

To configure SharePoint Server with increased diagnostic logging for authentication, follow these steps:

1. Launch **Central Administration**, click on **Monitoring**, and then under **Reporting**, select **Configure diagnostic logging**.

2. Expand **All Categories | SharePoint Foundation** and select **Authentication Authorization**:

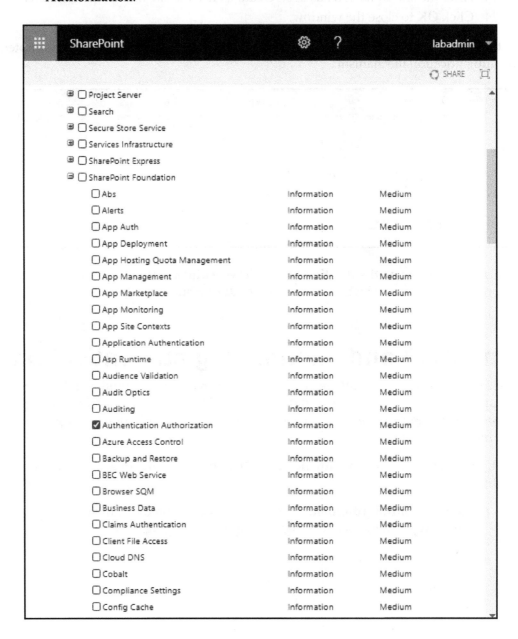

3. Under **Least critical event to report to the event log**, choose **Verbose**.
4. Under **Least critical event to report to the trace log**, choose **Verbose**.
5. Click **OK**.

Tracing events will be captured in the logs specified in the tracing folder. Event log data will be captured in the application log under the **SharePoint Foundation** source. To view the trace logs, you'll need to use the **Unified Log Service** (**ULS**) viewer from `https://www.microsoft.com/download/details.aspx?id=44020`.

When you have finished with the tracing and event log-capturing activities, return to the diagnostics logging area, clear any checkboxes that were set, and reset the log detail levels to report to the event log to **Information** and to **Medium** for the trace log.

Next, we'll look at some steps to troubleshoot authentication.

Troubleshooting authentication issues

When troubleshooting authentication, you'll likely need to follow the steps in the previous section to enable detailed logging and review the entries with the **ULS Viewer** application. You can search for the **Claims Authorization** or **Authentication Authorization** categories.

Depending on the types of errors discovered in the logs, it may also be a good idea to verify the authentication configuration for a web application or a zone. To verify or update the authentication configuration, follow these steps:

1. Launch **Central Administration**, and then go to **Application Management | Manage web applications**.
2. Select the name of the web application for which you're experiencing problems, and then on the ribbon, under the **Security** group, select **Authentication Providers**.
3. Select a zone to edit (all web applications have a **Default** zone; if you have additional zones, you should check through each methodically).
4. On the **Edit Authentication** page, verify (and update, if necessary) the settings for the claims authentication:
   - For Windows claims, ensure that **Enable Windows Authentication** and **Integrated Windows authentication** are selected and that either **NTLM** or **Negotiate (Kerberos)** is selected. If Kerberos is selected, remember that additional configuration steps are required. Select the **Basic** authentication option if it is required, and ensure that an SSL certificate is installed.

- For forms-based authentication, verify that **Enable Forms Based Authentication (FBA)** is selected and that the values for **ASP.NET Membership provider name** and **ASP.NET Role manager name** match the names specified in the web.config files for **Central Administration**, the web application, and SecurityTokenServiceApplication. For more information about editing these files, refer to https://docs.microsoft.com/en-us/previous-versions/office/sharepoint-server-2010/ee806890(v=office.14).
- For SAML-based authentication, verify that **Trusted identity provider** and the correct trusted provider name are selected.
- Under the **Sign in Page URL** section, verify the option for the sign-in page. If you are using a custom sign-in page for forms or SAML authentication, verify its URL. You may want to switch to the default sign-in page to determine whether the problem is with the sign-in page or some other part of the configuration.

5. Click **Save**.
6. Repeat the authentication attempt.
7. If the sign-in fails, review the ULS logs to determine whether the error messages are the same or different from this new authentication attempt.

Depending on the type of authentication being performed, you should also check the following items:

- Verify that the client web browser supports claims.
- If you are using Windows claims, verify that the client computer can access a domain controller (you can launch a Command Prompt and run nltest /dsgetdc: /force to force the discovery of an Active Directory domain controller).
- If you are using Windows claims, verify that SharePoint Server can communicate with Active Directory using nltest /dsgetdc: /force from a Command Prompt.
- If you are using forms-based authentication, verify that the ASP.NET role provider is configured correctly.

- If you are using SAML-based authentication, verify that both the federation provider (such as Active Directory Federation Services) and the identity provider (such as Active Directory) are reachable on the network.
- Consider using a network monitoring or tracing tool, such as Fiddler (`http://www.telerik.com`), to capture the authentication traffic:
    - For Windows-based authentication, check for traffic between SharePoint and the domain controller, as well as between the client computer and SharePoint Server.
    - For forms-based authentication, check for traffic web client computer and the SharePoint server, as well as the traffic between SharePoint Server and the server hosting the ASP.NET membership and role provider services.
    - For SAML-based authentication, check for traffic between the client computer and the federation endpoint, between the client computer and the identity provider (such as Active Directory), and the client computer and the SharePoint server.

These troubleshooting strategies will help you identify and resolve issues in your SharePoint authentication environment.

Summary

In this chapter, you learned about the different authentication methods available for SharePoint—Windows authentication, forms-based authentication, and SAML-based authentication. Each authentication method has its advantages, and its usefulness depends on both the client/server infrastructure (for example, using domain-based services and web applications for internal users) and the type of identity providers that are used (Active Directory, a database of some sort, or another organization's identity store). You also learned about the different monitoring and troubleshooting procedures to help you resolve issues.

In the next chapter, we'll learn about how to plan for and configure site collections in the SharePoint environment.

5
Managing Site Collections

In Chapter 2, *Planning a SharePoint Farm,* we looked at some basic concepts describing how SharePoint environments are composed: farms, built of individual servers hosting services, service applications, and site collections. While site collections are a way to organize content and resources, they are also a security boundary and control plane, meaning that we can set limits at that level that control what types of things can be configured or enabled in the sites associated with that site collection.

In this chapter, we're going to look at the core planning and deployment components related to site and site collection architecture:

- Planning and configuring the site collection architecture
- Planning for fast site creation
- Planning and configuring modern team and communication sites
- Planning and configuring modern lists and libraries
- Planning and configuring self-service site collections
- Planning and configuring site policies
- Reorganizing sites
- Deploying customizations
- Activating and deactivating site collection features

By the end of this chapter, you'll be able to articulate how you want to design your SharePoint Online site collection structure.

Planning and Configuring the Site Collection Architecture

Site collections and site hierarchies have been part of the SharePoint experience for quite some time. SharePoint Server and SharePoint Online share much of the same framework, so many of the design guidelines and recommendations are similar. However, SharePoint Online features one new structural paradigm that doesn't really exist fully on-premises: the hub site.

We can divide SharePoint architecture generally into two categories: classic and modern.

Classic SharePoint

The classic architecture is what we're familiar with; it has a curated look and feel and is typically static. Classic sites and site collections are built using a wide range of specialized templates, grouped into three families (Collaboration, Enterprise, and Publishing). The classic architecture is available in both SharePoint Server and SharePoint Online.

Let's look at the following diagram of a classic site architecture:

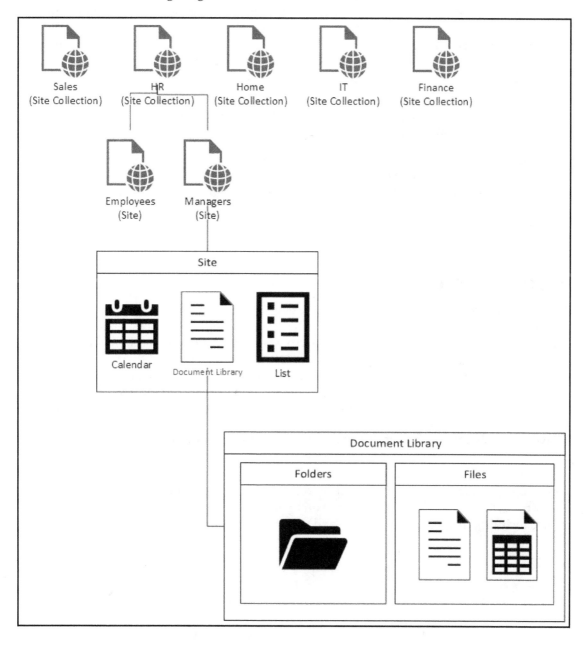

Content and applications are organized into sites, and then sites are further grouped into site collections, usually by business function, department, or agency. In classic SharePoint architectures, these can be arranged into logical hierarchies. Classic SharePoint sites have a rigid structure—a document's location is dependent on its placement inside the site and site collection hierarchy. For example, you may place a file called **Performance Review Template** in the document library on the **Managers** site, which is located inside the **HR** site collection. To navigate to it, you would have to navigate first to the **HR** site collection, locate the **Managers** site, expand the document library, and then expand the appropriate folder containing the document.

If a site needs to move from one site collection or hierarchy to another (for example, moving the **Employees** site from the **HR** site collection to the **Home** site collection), then a site migration must be performed. When this happens, all of the paths and URLs to documents and data stored in that site will change, causing bookmarks and file-sharing links to become invalid.

Modern SharePoint

The modern experience in SharePoint Online uses Office 365 Group-connected sites. In contrast to classic architecture, modern sites have no defined collection-site-subsite hierarchy. From a security boundary perspective, a modern site is essentially the same as a classic site collection. With the introduction of Office 365 Groups (and their integration into SharePoint), Microsoft also introduced a new SharePoint organizational concept: the hub site. While classic SharePoint sites and site collections are organized hierarchically, modern SharePoint sites are organized by association with a hub site.

However, in SharePoint Server, the Office 365 Group and SharePoint hub site constructs do not exist. As part of a modern design strategy (see `https://docs.microsoft.com/en-us/sharepoint/sites/plan-sites-and-site-collections`), Microsoft recommends using site collections as boundaries for work units and limiting the use of subsites. This will enable an easier transition to the SharePoint Online architecture. The other aspects of *modern site design*, including responsive templates and search, are the same.

Planning the site collection architecture

When planning your site collection architecture, you'll need to make some decisions about the design. These decisions include the following:

- **Web applications**: Microsoft design guidelines recommend one web application per farm for site collections and sites.
- **Security control**: From a security perspective, consider using separate farms for intranet versus public sites.
- **Site templates**: Once you know how many site collections you need, you can determine what kinds of site templates to use. For classic SharePoint architecture, this could include templates from the Collaboration, Enterprise, or Publishing site groups. For modern SharePoint architecture, the two most common templates are the **Communication site** and **Team site** templates.
- **Organizational design**: Try to use host-named site collections instead of path-named site collections; if your design has previously required alternate access mappings, use path-named collections instead. Only use alternate access mappings as a last resort, as the feature has been de-emphasized and host-named site collections are the design goal.
- **Storage**: When you purchase a SharePoint Online subscription (either standalone or as part of the Office 365 suite), your tenant is automatically allocated a pool of storage based on the number and type of user licenses purchased. Storage is automatically allocated by default, but you can further adjust or limit the storage as your organization requires.
- **Language support**: If your organization needs to support multiple languages, you may wish to use the **Multilingual User Interface** (**MUI**) feature to allow users to view sites and pages in a language other than the one configured for the site or site collection.

 MUI is not a translation tool; rather, it changes the display language for certain built-in default interface elements.

- **Governance**: Perhaps the most frequently overlooked planning task, governance is the process of determining things such as permission models, managing external user access, rights management, common user interface and design elements, the amount of customization you allow administrators to perform, how searches works, retention, and site life cycle management.

Site collections (and the sites they contain) also share common features, such as the following:

- Permissions
- Content types
- Web parts
- Navigation components
- Template galleries

When planning your SharePoint site collection architecture, be sure to evaluate the business requirements and map them to the design guidelines, taking care to understand the impacts the architecture has on the fluidity of the reorganization or how it might impact a future migration to SharePoint Online.

While Microsoft is guiding customers toward the new, modern SharePoint experiences, there may still be organizations that wish to stay with the SharePoint classic architecture for a while due to application or design dependencies or other organizational change issues. Therefore, there will still be a need to determine site collection and subsite structures.

The topic of design, however, does still have content applicable to modern site architectures as well. As previously mentioned, Microsoft has recommended using the design principle of Office 365 sites as boundaries for SharePoint Server modern sites. Building site designs that can be reused throughout the organization will help you achieve a level of consistency.

Determining who will use the site

As discussed previously in this chapter, with SharePoint classic architecture, this question may be answered by creating site collections per business unit, locality, or other business-specific concepts. You may create a collection with a large scope, such as a region (US, Europe, or Asia), and then create sites inside those site collections for business units, products, departments, or agencies.

The SharePoint modern architecture works similarly—but instead of creating rigid site structures, you can create a flat site architecture. In SharePoint Online, you can designate hubs, and then associate sites with those hubs. With SharePoint Server, you'll join the sites together with common navigation elements and site design templates. Since each site is its own security boundary, you're able to move sites to new hubs without accidentally compromising access controls.

Access to classic site collections is managed using the site permissions control for the collection, and by default, sites and subsites inherit the permissions settings of their parent sites. This is the direct opposite of the modern SharePoint permissions experience, where access to each site is governed by its own unique site permissions (typically specified when the site is created). Since the modern SharePoint experience works on an opt-in basis (that is, site owners or creators must choose their members), it is less likely that a member will get access to content that they shouldn't.

Determining the content of the site

A SharePoint site can host curated and collaborative content as well as applications. Understanding the types of templates available and add-ons or applications that will be installed will help to determine what type of content will be stored on a site. The diagrams and tables located earlier in this chapter show you what kinds of templates are available to use in your design.

Planning the navigation structure

Depending on the type of site structure you use (classic or modern), you may have access to structural or managed navigation. Evaluating what type of site organization and grouping you will use and which site experience (classic or modern) will determine what types of navigation options you have available. Managed navigation is only available for classic sites.

Microsoft is moving toward modern site experiences because of their flexibility and rendering optimization on multiple devices, so those navigation designs should be given preference.

Determining access to sites and content

Classic site collections and modern sites are functionally equivalent in terms of the access controls available. As we mentioned previously, a modern site is a site collection with a single site. One of the most important decisions to make from a security standpoint in either design scenario is managing access. If you use classic architecture (a site collection with one or more sites and each possibly containing multiple subsites), you'll likely take advantage of some level of permission inheritance. If you use modern architecture (a one-to-one mapping of sites and site collections), you'll be managing access at a higher level, making discrete decisions about direct access to site content.

Next, we'll look at what you can do to take advantage of the SharePoint Server fast site creation feature.

Planning for Fast Site Creation

SharePoint Server 2019 includes a feature called **fast site creation**. Fast site creation allows the provisioning of new sites in a matter of seconds. Fast site creation is only enabled with the following templates, where POLICY# is the identity of the site template being used:

- A OneDrive personal site SPSPERS#10
- A team site (modern only) STS#3
- A communication site SITEPAGEPUBLISHING#0

Fast site creation is activated when using the following provisioning mechanisms:

- OneDrive personal site auto-provisioning
- The **Create Site** button in SharePoint Home
- Using the -CreateFromSiteMaster parameter with the New-SPSite PowerShell command

There is nothing to do to further enable fast site creation—you just need to use the correct SharePoint site templates and provisioning methods.

Planning and Configuring Modern Team and Communication Sites

In the previous sections, we discussed some of the design principles around classic and modern sites. Microsoft's design recommendation going forward is to build modern sites, so we'll focus on that as well.

Creating a modern site (referred to as a site collection in classic architecture) is quite simple from an admin interface. In this example, we'll use the **Central Administration** interface (if you've configured self-service site creation, you'll be able to do it via that route as well).

Use the following steps to complete the task:

1. Navigate to **Central Administration**, and then select **Application Management** | **Create site collections**.
2. Enter a title, description, and URL.
3. Under **Template Selection**, if you want to create a modern team site, choose the **Collaboration** tab and then select **Team site**. If you want to create a modern communication site, choose the **Publishing** tab and then select **Communication site**:

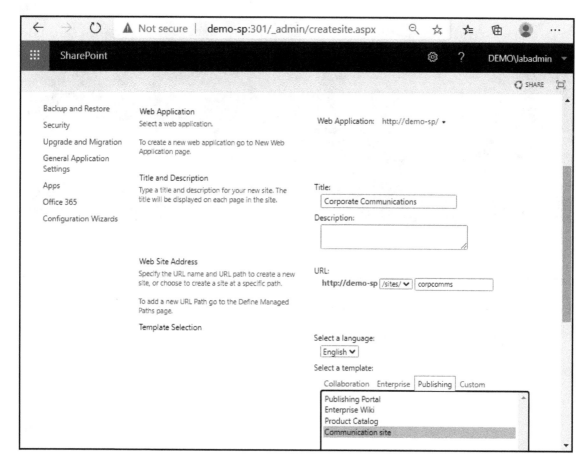

4. Specify a primary site administrator and click **OK**.

Navigate to the site to ensure it has been created successfully. Modern site templates have a responsive design and should resize well. You can see an example of the modern **Communication site** template in the following screenshot:

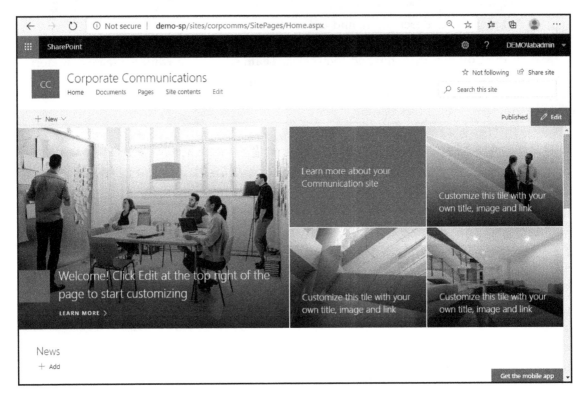

From here, you can continue designing it or begin adding content.

Next, we'll look at configuring modern lists and libraries.

Planning and Configuring Modern Lists and Libraries

The modern list and library experiences were first introduced to SharePoint Online in 2016. They have since been introduced to on-premises versions of SharePoint.

The modern list and library experiences are designed to make it easier for users to navigate and update lists, as well as upload and interact with documents.

Modern lists

Modern lists are easily added from a modern site's **Site Contents** menu. Simply click on the **+New** button at the top of the page and select a list:

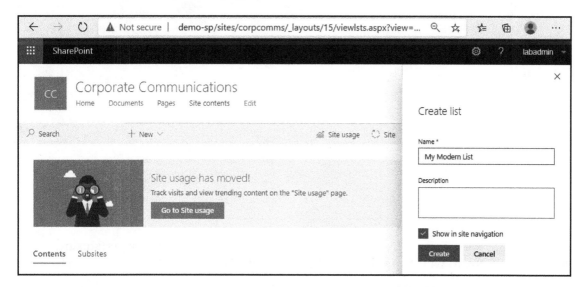

Selecting the **Show in site navigation** checkbox will cause the newly provisioned site to show up on the navigation bar where the **Home**, **Documents**, **Pages**, and **Site contents** links are displayed.

Modern lists allow users to easily add columns, as well as search, filter, and export data, as shown in the following screenshot:

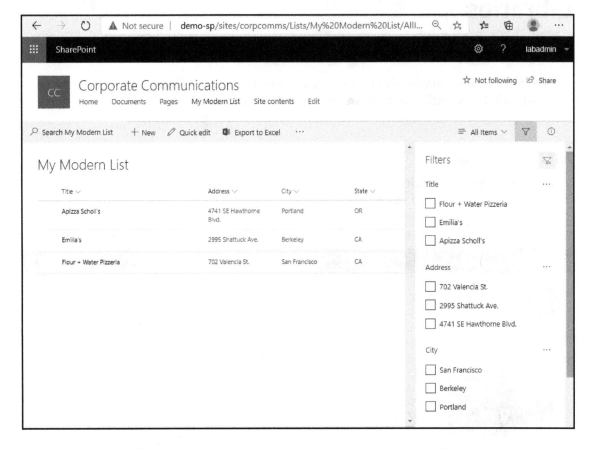

By expanding the ellipsis in a row item, you can learn about the compliance status or details of an item, as well as use it to start a workflow.

Next, we'll look at modern libraries.

Modern libraries

The look and feel of modern libraries are modeled after the new OneDrive interface. Like a modern list, the easiest way to create a modern library is to use the **+New** link on the **Site Contents** page. After creating a library, you can navigate to it. As you can see in the following screenshot, the most common activities appear on the menu bar, and you can engage in similar filtering and searching activities as you can with a list:

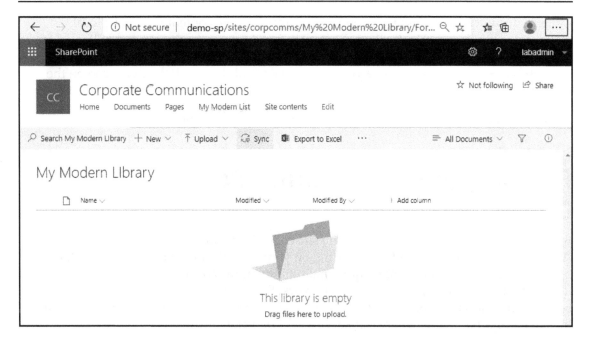

The new user interface does not prevent you from accessing the controls, however. You can click the gear icon at the top of the page and then select **Library settings** to navigate to some of the familiar controls, such as changing versioning, audience targeting, and enterprise keywords:

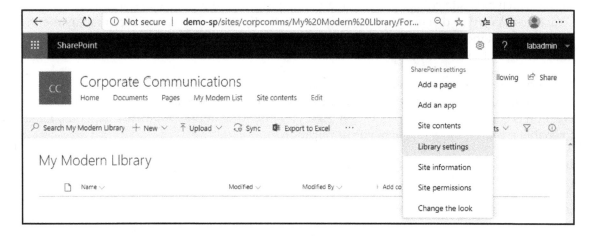

Modern document libraries also support synchronizing file content with your local computer using the OneDrive client. To initiate or set up synchronization, you'll need to install the latest version of the OneDrive client (`https://go.microsoft.com/fwlink/p/?LinkId=248256`), sign in to it using your credentials, and then click the **Sync** icon on the library ribbon.

In the next section, we'll discuss self-service site collections for SharePoint.

Planning and Configuring Self-Service Site Collections

By default, only administrators can create site collections (through either Central Administration or PowerShell). Depending on your organizational needs, this may be desirable. For organizations that require maintaining tight governance and control over how content is managed, this method may work. Some organizations, however, may benefit from allowing more dynamic provisioning of resources (such as with short-term or informal projects). This is where self-service site collection creation can help.

Planning

The self-service site collection administration functionality allows you to define a path under which new site collections are created, helping you to organize user-created structures. As we've discussed previously, site collections are atomic units of management, meaning that whoever owns the site collection will generally be responsible for administering the membership and access controls.

Self-service site creation can also be configured to create sites inside a site collection. Sites, by default, inherit the permissions of the parent site collection. Self-service site collection creation is enabled per web application, so if you intend to deploy it farm-wide, you'll need to enable it individually on each web application.

It's important to note that self-service site creation can only create path-based collections (not host-named collections).

Important planning considerations include the following:

- Which web application(s) will have self-service site creation enabled?
- Which, if any, retention configurations should be deployed?
- How will you handle unused sites?

- What type of storage quotas or templates will be used?
- Will there be secondary site contacts in the event that the primary contact leaves the organization?

Once those determinations have been made, you can continue on to configuring self-service site collection creation.

Configuring self-service site creation

Configuring self-service site creation for SharePoint Server 2019 is a relatively straightforward process.

 Hybrid self-service site creation is another method of self-service site creation. It redirects the requests to SharePoint Online. This is discussed in `Chapter 12`, *Implementing Hybrid Teamwork Artifacts*.

To configure self-service site creation, follow these steps:

1. Launch **Central Administration**, select **Application Management**, and then select **Manage web applications**.
2. Select the web application for which you wish to enable self-service site creation, and then, under the **Security** group, click **Self-Service Site Creation**:

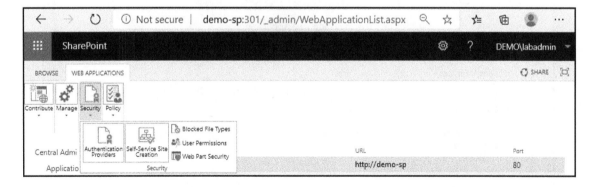

3. Under **Site Collections**, select the **On** radio button to enable self-service site creation. You can further select the web application under which new sites will be created, a path, and your ownership information for the site:

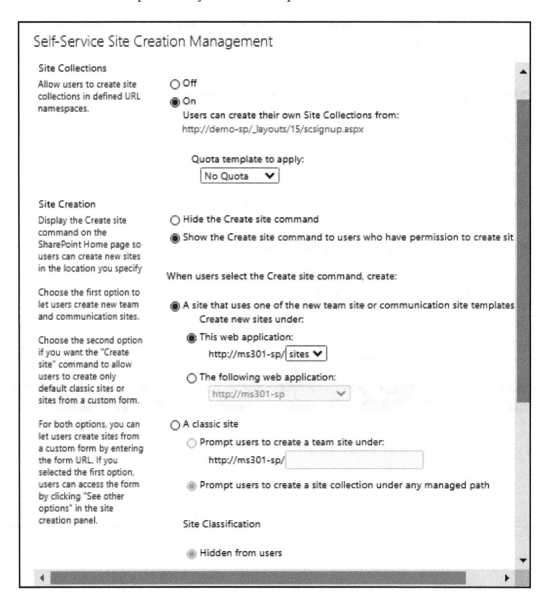

4. Click **OK** when finished.

If you enable the **Show the Create site command to users who have permissions to create sites** radio button, the **+New** button should appear for users who have the relevant permissions. Otherwise, you can direct users to `http://<webapplication>/_layouts/15/scsignup.aspx`. Users should see a dialog similar to the following:

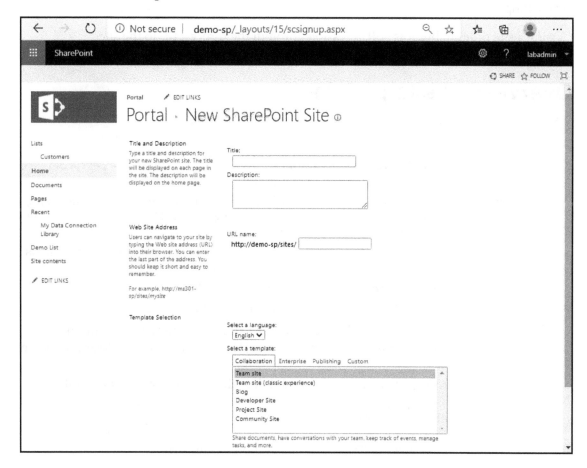

Users can go through the process to provision their own sites.

Next, we will learn about site policies.

Planning and Configuring Site Policies

Site policies are a method to help manage the governance of the life cycle of sites. A site policy can be used to determine when a site will be closed and when it will be deleted. Closing a site closes or deletes any subsites as well; likewise, if an Exchange mailbox has been configured for the site, then this is deleted as well.

Closing a site hides it, but it is still accessible through the direct URL to the site.

Overview of site policies

A site policy specifies the conditions under which to close or delete a site. There are four configuration options:

- **Do not close or delete the site automatically**: This is essentially *no policy*. Site owners must close and delete sites manually.
- **Delete the site automatically**: This setting specifies that if an owner closes a site, it will be deleted automatically according to the settings. The delete policy specifies a rule with the following options:
  - Triggers and timing for deletion, such as "2 months after the site is closed" or "1 year after the site is created"
  - Email notifications about pending deletions
  - Whether owners can delay or postpone site deletion
- **Close the site automatically and delete the site automatically**: An automated way to manage the life cycle of the site that takes care of when to close and delete sites.
- **Run a workflow to close the site, and delete the site automatically**: Similar to the previous option, it allows you to govern a site's creation and life cycle options, with the addition of running a workflow.

With those options in mind, let's look at how to configure a site policy.

Configuring site policies

Site policies are created at the site root of a site collection and are available for every site collection on a farm. To define a policy, follow these steps:

1. Navigate to a site collection's **Site Collection Administration** menu by clicking on the gear icon, clicking **Site Settings**, and then locating **Site Policies** under the **Site Collection Administration** section:

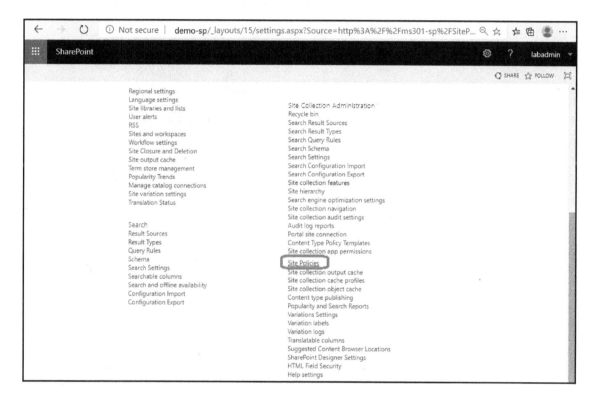

2. Click **Create**.

3. Fill out the policy value fields, including **Name**, **Description**, closure and policy guidelines, and whether to mark a site as read-only when it's closed:

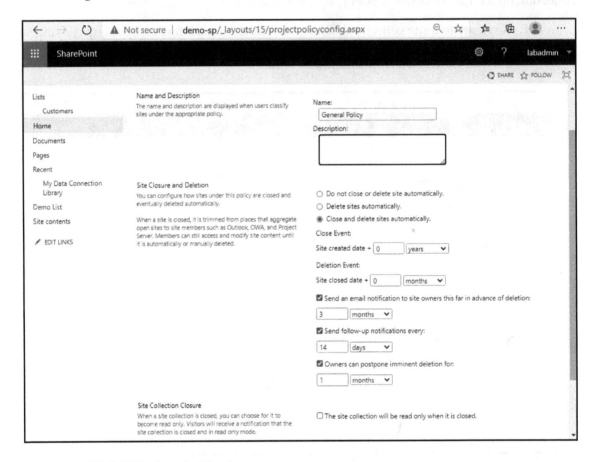

4. Click **OK** when finished.

The new policies will take effect when the timer jobs for site closure and deleted sites run out.

Next, we'll look at methods for moving or reorganizing site content.

Reorganizing Sites

As we discussed in the *Planning and configuring site collection architecture* section, Microsoft recommends using the one-to-one, site-to-site collection architecture going forward to help make it easier to transition to SharePoint Online. In SharePoint Online, since modern sites all use top-level site designations, there's no need to worry about breaking file sharing links or absolute paths to documents and resources due to organizational restructuring.

If you use classic architecture designs for sites, you may find that you need to move content based on how your business operates (for example, if a product in the development phase moves to production and is now owned by a different department). There are a few options available, such as using the **Save site as template** method or the **Site Content and Structure** administration options.

Saving a site as a template

The **Save site as template** method will save your site as a template that can be redeployed. It does have a default limitation of 50 MB, though, so if your site has a lot of content, this method will not work.

 You can increase the site template output size to a limit of 500 MB, but for large sites, this limit won't be enough. To increase the limit, launch an administrative prompt, and run the following command:
```
stsadm -o setproperty -pn max-template-document-size -pv
524288000
```

Follow these steps to save a site as a template:

1. Navigate to the top-level site of the collection.
2. Click the **Settings** gear icon, and then select **Site Settings**.
3. Under **Site Actions**, click **Save site as a template**.
4. Enter a name and description, and then select the box to save and include site contents.
5. Click **OK**.
6. Click **Solutions Gallery** to navigate to the solutions gallery so that you can download the file if you want a backup.
7. Click the name of the solution you wish to download (specified previously in *step 4*), and then click **Save**.
8. In the **Save As** dialog box, browse to a location to save the file, and click **Save**.

Later, to load the content, we do the following:

1. Navigate to the top-level site of the new location and create a site.
2. Under the **templates** section, select **Custom** and then select the template you saved previously.

The site has been imported.

Site Content and Structure

If you have more content, you may want to use the **Site Content and Structure** capability. **Site Content and Structure** requires the **Publishing** features to be activated. Activating them will disable the **Save site as template** functionality.

To access the **Site Content and Structure** interface, navigate to `http://<site>/_layouts/sitemanager.aspx`. You'll see a Windows Explorer-like browser that shows the entire site structure:

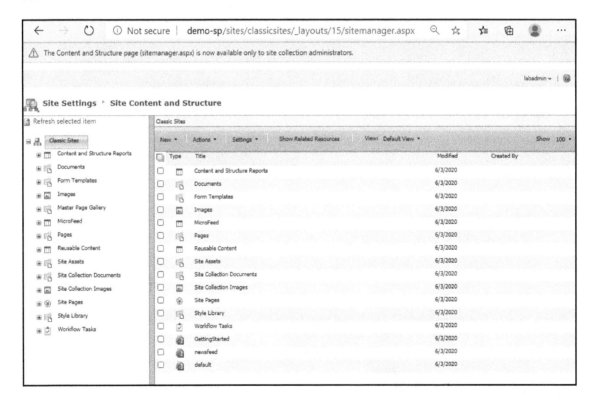

Expand a folder, such as **Documents**, and then select the items to move. Click **Actions**, and then select either **Move** or **Copy**:

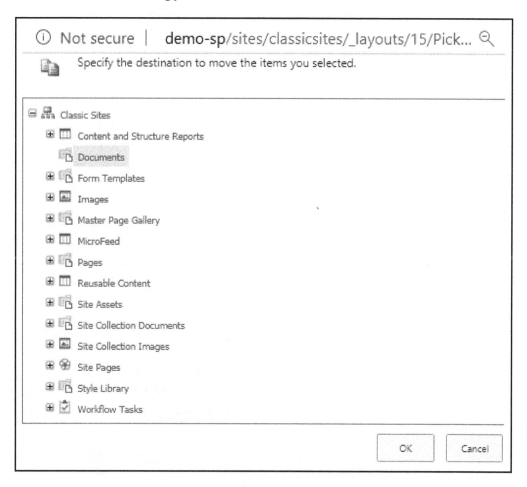

Something important to note about the **Site Content and Structure** feature is that it can only be used to move content within the same site collection (it can't be used across site collections).

If you need to move in excess of 500 MB (after customizing the **Save site as template** option) across site collections, the best choice is likely a specialized migration tool.

Deploying Customizations

As a web platform, SharePoint Server is customizable. The look and feel of sites and collections can be standardized across your deployment using concepts such as **Cascading Style Sheets (CSS)** and master pages. When performing extensive customization of the SharePoint experience, you should spend time looking at the SharePoint development patterns and practices at `https://docs.microsoft.com/en-us/sharepoint/dev/solution-guidance/sharepoint-development-and-design-tools-and-practices`.

From a branding perspective, there are some key terms and concepts with which you should become familiar:

Terminology	About
Alternate CSS	A CSS file other than the default that you can apply to the look and feel of your site.
Composed look	A combination of fonts, palettes, a background image, and an associated master page that can be applied to a site.
Content Search Web Part	A web part used to display content from search results.
corev15.css	The main CSS file that contains most of the layout and design for SharePoint. The `_layouts\15` folder is the default file location.
CSS	CSS provides detailed instructions to a browser on how to render elements on a page, such as fonts, spacing, and colors.
CSSRegistration	A reference in a master page that loads most CSS that is applied to most of the default user interface.
Custom action	Actions you can use to customize the interactions of the ribbon.
Device channels	A method of targeting published content to render differently depending on a device.
Display templates	Templates used by Search web parts to show the results of a query made to the search index.
Image rendition	Provides the capability to resize publishing site images differently.
Managed metadata	Structured terminology or taxonomy that can be used to categorize data and be used as the basis for managed navigation.
Managed navigation	Navigation for publishing sites built based on metadata.
Master page	Standardizes the behavior and presentation of the navigation areas of a SharePoint page.
Master Page Gallery	A special document library in SharePoint where branding components such as master pages, layouts, JavaScript files, CSS, and images are stored by default.
Page content control	A control on a publishing site where a web part can be added.
Page layout	A template for a SharePoint publishing site page that lets users lay out information on the page in a consistent way.
Quick Launch	Manages the navigation elements on the left side of the page of a collaboration site.

REST	A stateless architectural style that abstracts architectural elements and to interact with pages.
Root web	The first web page in a site collection.
Seattle master	The default master page for SharePoint team sites and publishing sites.
Structured navigation	A navigation structure for publishing sites that is based on the site hierarchy.
Theme	A simple way to apply branding to a SharePoint site.
Theming engine	Files and other artifacts used to define the look and feel of composed looks.
User Agent String	Configuration data submitted by the browser to identify its capabilities.
User Custom Action	A CSOM property that returns the collection of custom actions for a website, list, or site collection. The default file location is 15\TEMPLATE\FEATURES.

While developing detailed customization packages is beyond the scope of this book, you should familiarize yourself with the terminology used in developing the SharePoint user experience.

Next, we'll look at the basics of activating and deactivating features.

Activating and Deactivating Site Collection Features

Most site collection features are disabled or turned on in SharePoint by default. Depending on the type of site (classic or modern) and the features you want to deploy, you may need to turn them on. This is achieved through the **Site settings** applet on the site. To enable or disable features, follow these steps:

1. Navigate to the **Site settings** page:
   - On modern communication sites, select **Site contents** in the top menu bar and then click **Site settings**.
   - On a modern team site, select **Site contents** in the left pane, and then click **Site settings** in the top navigation bar.
   - On classic sites, click **Settings** and then click **Site settings**.
2. On the **Site settings** page, click **Site collection features** under **Site Collection Administration**.
3. Do one of the following on each site collection feature you want to enable or disable:
   - Click **Activate** to enable the site collection feature.
   - Click **Deactivate** to disable the site collection feature.

Some features have dependencies on other features, either at the site or site collection level. If you encounter a feature that has dependencies, you may receive a message indicating you need to enable another feature first.

Summary

In this chapter, we've covered some of the core SharePoint Server site collection planning concepts, including choosing a site design architecture (classic versus modern), interacting with modern site components, such as lists, libraries, and site templates, and managing features.

From an architectural perspective, Microsoft's recommendations are to focus on modern design concepts, paving the way for migration to SharePoint Online.

We also covered site policies to help manage the governance of sites, and to move content between sites using the **Save site as template** method and the **Site Content and Structure** tools.

In the next chapter, we will dive into Business Connectivity Services for interacting with remote data.

6
Configuring Business Connectivity Services

Business Connectivity Services (**BCS**) solutions are custom solutions that integrate with SharePoint Server (as well as SharePoint Online). Developing a BCS solution requires both deep business application knowledge (such as what types of entities need to be managed) and development experience with Visual Studio and SharePoint Designer or the Microsoft Office Developer Tools.

BCS can be used in a number of scenarios, such as the following:

- Retrieving and presenting external data from enterprise applications or web services
- Enabling offline use of external data
- Connecting structured data from enterprise applications with unstructured data such as documents and images
- Building applications to update and interact with data in enterprise or external applications

While building external applications that leverage BCS is outside the scope of this book, it's important to know what the capabilities of the system are and how to set it up, since the application developers in your organization may need to use these capabilities.

Microsoft's forward-looking stance is to look at using Power Platform-based solutions for new development efforts. Additionally, for scenarios that only require reading data, a data gateway to connect to Power Platform may also be a suitable solution. You'll learn more about data gateways in Chapter 14, *Implementing a Data Gateway*.

In this chapter, we'll cover the following topics:

- Overview of BCS
- Overview of Secure Store

- Planning for external content types
- Configuring connections to external data sources
- Planning and configuring search for BCS applications

By the end of this chapter, you'll have an understanding of basic BCS concepts.

Let's go!

Overview of BCS

As mentioned previously, the purpose of BCS is to provide a mechanism to retrieve and manipulate data stored in external business applications or systems. BCS can be used to embed remote datasets in your own applications, giving users a familiar interface to manipulate or view data.

For example, if you are a reseller of restaurant supplies, you may want to connect to the manufacturer or other wholesale distributors and import their data catalogs into custom applications that your employees use to quote and resell equipment to end customers. Using BCS, you can create connections to your suppliers' applications and render that data in your own applications.

Depending on the data source and external data configuration you need, BCS can be used to support **create, read, update, delete, query (CRUDQ)** operations.

Overview of Secure Store

Secure Store can be thought of as a type of credential vault for SharePoint. You can store credentials for authenticating into external applications. Credentials are stored in a configuration item called a *target application*.

Before the Secure Store service can be used, however, you'll need to prepare it to be able to store and encrypt credentials.

To prepare Secure Store, follow these steps:

1. Launch Central Administration.
2. Select **Application Management**, and then select **Manage service applications**.

3. Select **Secure Store Service** and then click **Manage** from the ribbon:

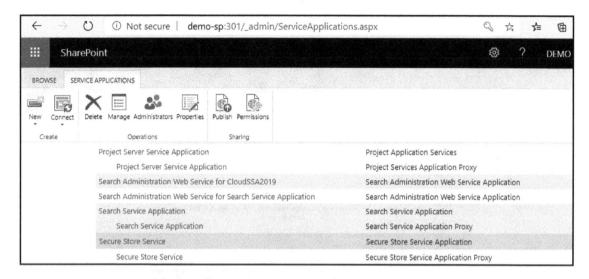

4. The first time you manage the Secure Store Service application, you'll be prompted to generate a key, as shown in the following screenshot:

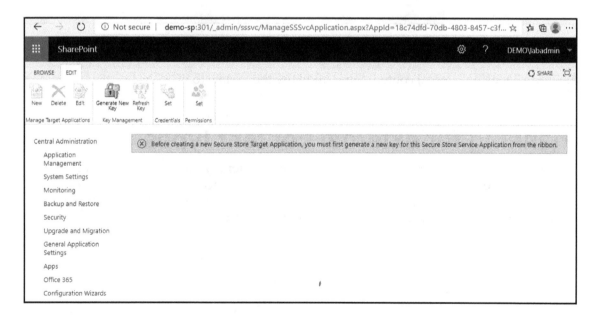

5. On the ribbon, in the **Key Management** group, click **Generate New Key**.

6. Enter a passphrase to encrypt the database. Record the passphrase in a safe place, as it will be not possible to retrieve it in the future. Click **OK**:

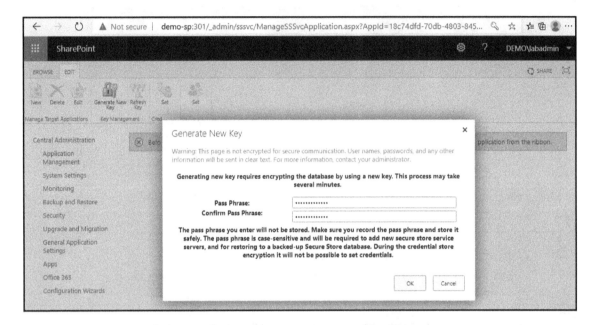

Once the key has been generated, the Secure Store service application is ready to use.

Next, we'll look at external content types and how they relate to BCS.

Planning for external content types

BCS utilizes *external content types*. Similar to normal content types (which can be metadata columns, properties, and workflows that are applied to a piece of data), an external content type allows you to apply those same concepts to external data. External content types allow the reuse of metadata and actions on external data sources.

External content types can be used to create and represent external data as the following types of objects:

- **External lists**: External lists are the most common type of object in BCS solutions. External lists use an external content type as their source, allowing you to create a SharePoint list populated with external data. External lists function just like any other SharePoint list, providing a familiar interface to interact with data. External lists, depending on the connected system, may allow you to write back to the source. If configured, users can make changes to content in the external list, which will get synchronized back to the external data source.
- **External data columns**: Similar to external lists, external data columns allow users to link external data columns to existing SharePoint lists.
- **Business data web parts**: In addition to using building blocks such as external lists and data columns, you can also use web parts, such as Business Data List, Business Data Item, Business Data Related List, and Business Data Actions.
- **External content type picker**: The external content type picker can be embedded in pages or forms to give users a way to choose data items from an external data source.
- **External item picker**: Similar to the external content type picker, the external item picker is used to provide the external data picking capability from within a custom app.
- **Profile pages**: A profile page displays the details and properties of an external item.
- **Custom pages**: Here, you can use the SharePoint object model, client object model, and REST interface to programmatically access external data.

Knowing how the external content will be used is required if you want to successfully choose the appropriate content types.

After selecting what data you'll use and how you'll use it, you can select the *external content type* that meets those requirements. Then, you can begin configuring BCS accordingly, as we'll see in the next section.

Configuring connections to external data sources

Once your organization has decided on what external data sources to interact with, you can work with your partners, suppliers, vendors, or customers to create a connection to their data source.

When connecting to external applications or data sources, you'll need to work with the provider of the application or data source to obtain some sort of credential, as well as any other special configurations that are necessary to successfully configure the relationship. In this example, we're going to configure BCS to connect to Salesforce to give you an idea of how these components work together.

As a prerequisite, you'll need to configure a Salesforce-connected app. Configure OAuth settings and grant the API permissions to access and manage data. Record the consumer key and the consumer secret.

 Note: This example is really intended to show one way that BCS can be used at a basic level. For information on configuring Salesforce, please see the Salesforce developer documentation.

In the next few sections, we'll configure a Secure Store target application, which will allow BCS to authenticate to the external data source. Then, we'll configure BCS so that it can connect to an external data source.

Configuring Secure Store target applications

Before we can configure connectivity to an external application, we need to configure the Secure Store service application so that it can host credentials. Without a secure store, BCS will have no way to log in to the remote data source.

To configure Secure Store, follow these steps:

1. Launch Central Administration, navigate to **Application Management**, select **Manage service applications**, and then select the Secure Store service application.
2. On the ribbon, in the **Manage Target Applications** group, select **New**:

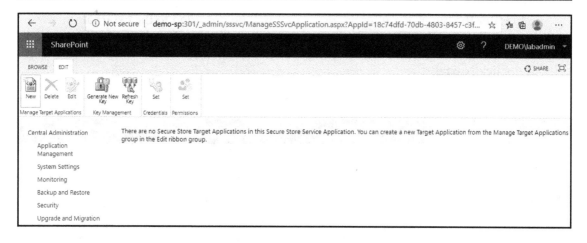

3. Enter values for **Target Application ID** and **Display Name**, as well as a contact email address for the user or group who will be responsible for administering or maintaining this credential. As per our application configuration, we're going to select a target application type of **Group**, but your application's requirements may differ. When finished, click **Next**:

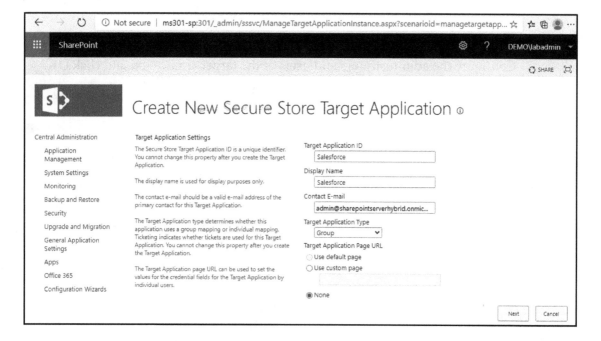

4. Enter the types of credential fields that will be provided to the external application. For this particular application, we'll select **Username** and **Password**, but the configuration will depend on the requirements of the remote application. Click **Next**:

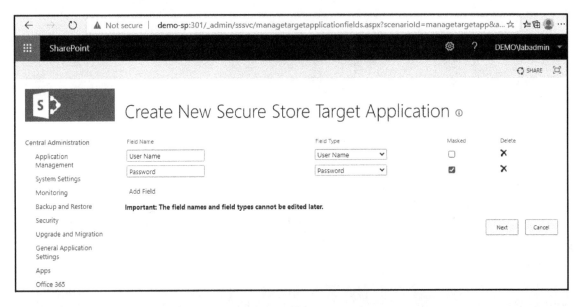

5. Configure the administrators and users of the target application and click **OK**:

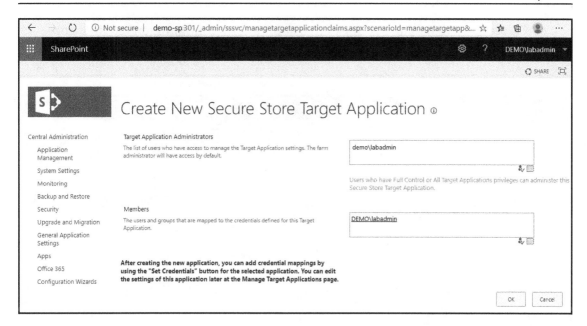

6. After the target application has been created, select it from the list and click **Set** in the **Credentials** group of the ribbon:

7. You'll need to follow your external application's configuration requirements here. In this example, we're providing a credential to the Salesforce instance:

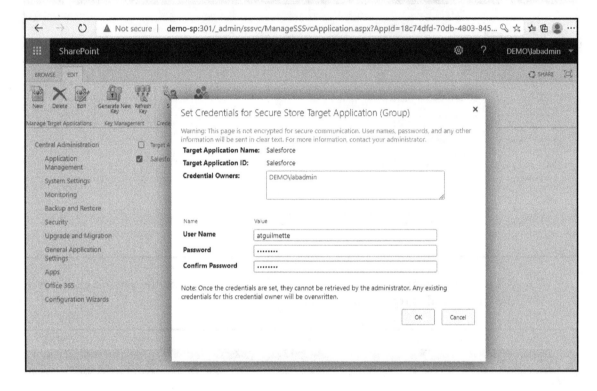

8. As per our connected data source's instructions, we need to repeat this process again, only this time, we're going to store the **Consumer Key** and **Consumer Secret** objects:

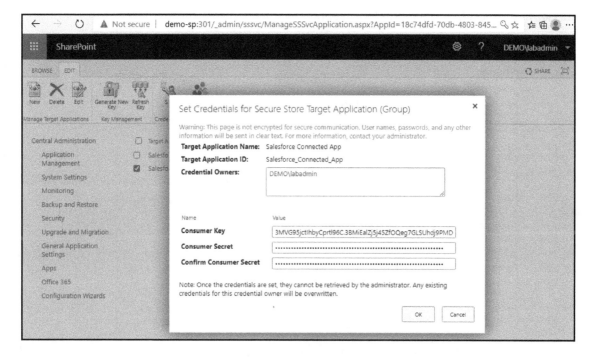

The target application credentials have been saved. Next, using the tools provided by the sample application, we'll import a sample data model.

Importing the BCS model

Data is represented in BCS through a *data model*. The data model file is formatted as an XML file, with nodes and elements related to the remote data structure. An example data model file is shown in the following screenshot:

To import a model into BCS, follow these steps:

1. Launch Central Administration, select **Application Management**, select **Manage service applications**, and then click **Business Connectivity Services**.
2. In the **Permissions** group of the ribbon, select **Set Metadata Store Permissions** and add your account. Select the permissions you want to use and click **OK**.
3. On the ribbon, click **Import**:

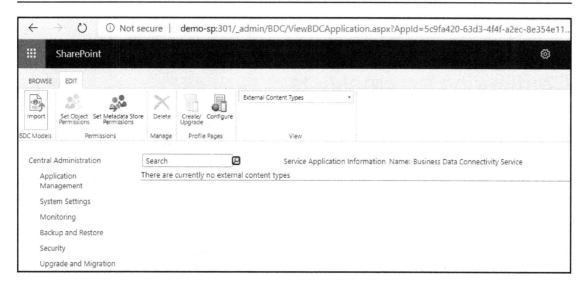

4. Browse to the file you wish to import and select the type. In this example, we're importing a model file. Click **Import**:

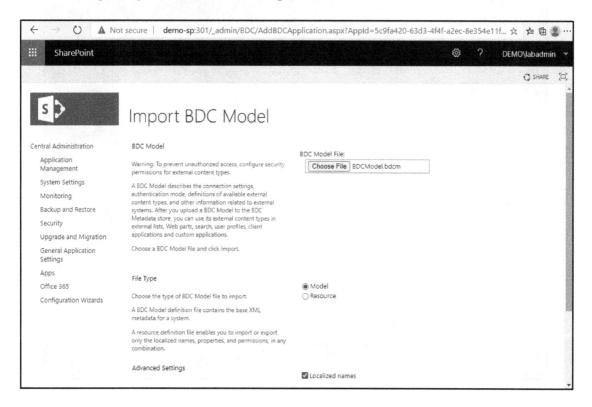

Once the BDC model has been imported, you can begin creating external lists.

Configuring external lists

External lists will allow users or applications to query or otherwise manipulate the data represented by the BDC model. External lists and content types can be defined with SharePoint Designer. To configure an external list, follow these steps:

1. Navigate to a SharePoint site and select **Site Contents**.
2. Click **New** | **App**.
3. Select **External List**.
4. Fill out the properties of the external list. You'll need to give it a name and then select a source external content type. In this case, we're selecting an entity that we imported from a BCS model file. Click **Create**:

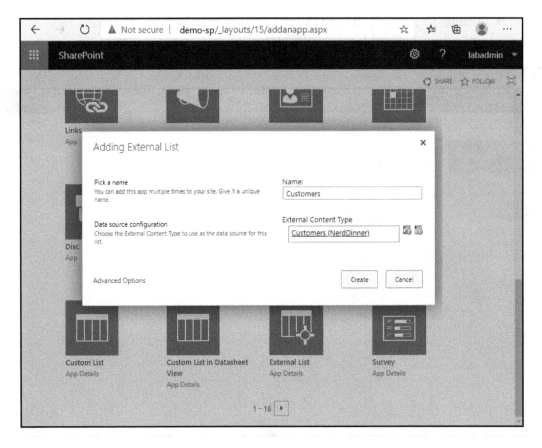

5. Refresh the **Site Contents** page, and then click on the newly created list. Note that the item has an **External list** instance displayed in the **Type** column:

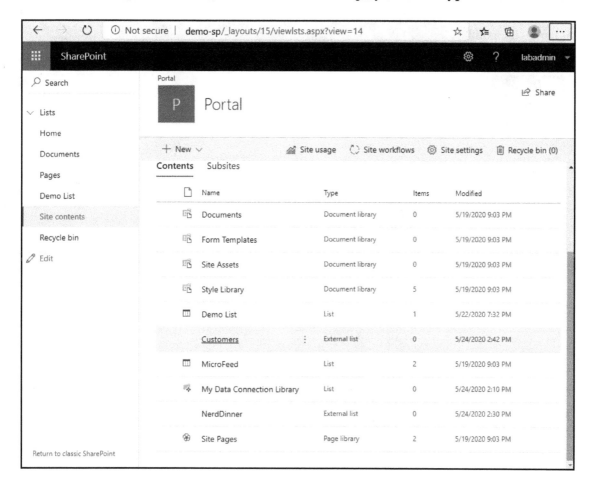

6. Verify the contents to make sure it has retrieved the external data:

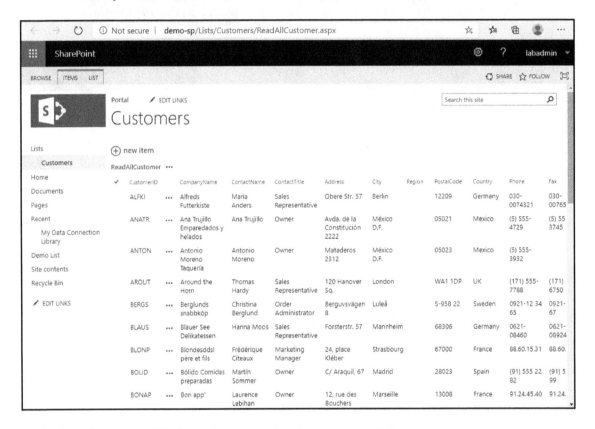

With that, the external list's configuration has been successfully completed.

Planning and configuring search for BCS applications

Since an external list functions like any other SharePoint content list, it can be indexed and searched. The content isn't available to search by default, however, so it must be added as a content source. Once you've added the BCS data as a content source and crawled the BCS content, you will be able to search for it.

To configure search for BCS data, follow these steps:

1. Launch Central Administration, select **Application Management**, select **Manage service applications**, and then click **Search**.
2. Under **Crawling**, click **Content Sources**.
3. On the **Manage Content Sources** page, click **New Content Source**:

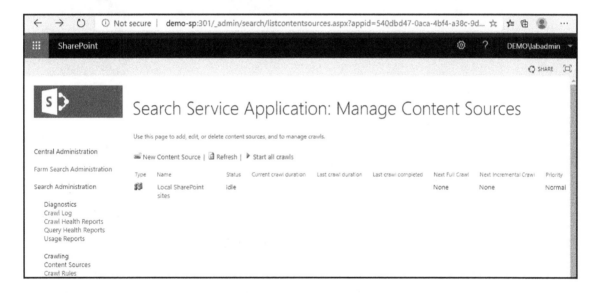

4. Enter a name for the content source and select **Line of Business Data** as **Content Source Type**. Select a **Business Data Connectivity Service Application** option and then select a data source. Click **OK** when you're finished:

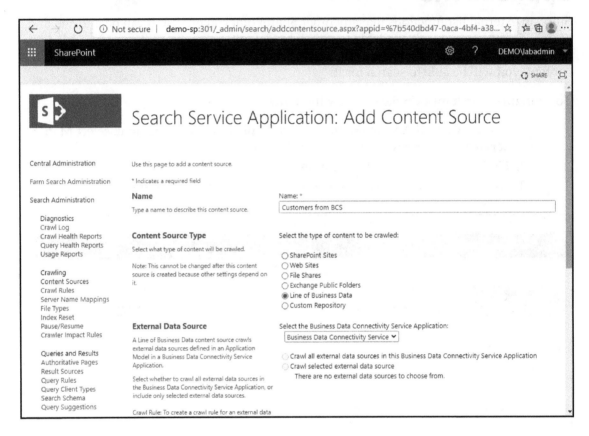

5. Right-click the data source and start a full crawl.

After you've done this, you can search for content that has been indexed from the external list. The availability of the content to be crawled will depend on the data source and the model file you imported earlier.

Summary

In this chapter, we covered BCS and Secure Store, and how they interact. BCS can be used to render remote datasets in the local SharePoint system, making it available for queries or to custom applications. Depending on the remote data, BCS can also support data modification, delete, and query operations. You learned how Secure Store, credential storage, and management services can be used by BCS when connecting to remote data sources. Finally, we walked through some sample configurations with both BCS and Secure Store.

In the next chapter, we'll cover managed metadata concepts and how metadata can be used to improve the discoverability of data.

7
Planning and Configuring Managed Metadata

In this chapter, we will discuss one of the more powerful aspects of SharePoint—metadata. We're all familiar with the information we put into SharePoint, which takes the form of documents, spreadsheets, images, and so on. Metadata is detailed information about these documents and artifacts.

For example, a document might be about a particular business process, but the metadata about the document will include things such as the author's name, their department, the date the document was created or updated, the subject of the document, and keywords and phrases that we might use when searching for the document. Organizing and managing metadata will help your users to locate the resources they need more efficiently.

Configuring and using managed metadata brings a lot of advantages to a SharePoint environment:

- **Consistency**: Using managed metadata and term sets allows users to select from predefined terms, ensuring consistency in spelling and phrasing.
- **Improved discoverability**: By using consistent names and terminology across your enterprise, you improve the user's ability to search for content and refine results.
- **Metadata-driven navigation**: Metadata can be used to create navigation elements for sites based on terms, as well as to create dynamic views of information in libraries and lists.
- **Increased flexibility**: Because of the ease of updating term sets, you can quickly update the classification of items across an organization. For example, if a business updates or changes a product name, administrators can merge terms in a term set. Alternatively, if similar terms are discovered, administrators can merge the related terms. In either of these cases, all of the content previously tagged with either of the terms is updated.

The topics covered in this chapter are as follows:

- Planning and configuring Managed Metadata Service applications
- Planning and configuring Managed Metadata Service features
- Planning and configuring Term Store security
- Planning and configuring Term Store structure
- Planning and configuring Term Store languages
- Maintaining the Term Store

By the end of this chapter, you'll be able to successfully configure a Managed Metadata Service implementation. We have a lot of information to cover in this chapter, so let's get started!

Planning and configuring Managed Metadata Service applications

In this series of sections and tasks, you'll learn about the terminology related to metadata and content classification, as well as the tools and options available for making metadata available to your users. Planning a successful Managed Metadata Service implementation requires input from both business and technical resources to ensure that there's a well-reasoned structure that takes into account the taxonomy and terminology that the organization uses.

Overview and terminology

When discussing metadata, the following terms are used to describe the core concepts:

- **Taxonomy**: This is a formal, hierarchical grouping of words, labels, and terms used to describe something.
- **Folksonomy**: This is an informal classification system containing labels, keywords, and terms applied by end users to describe something. If you've seen blog sites that have categories and tags (such as a tag cloud), that is an example of a folksonomy.

- **Terms**: This is a specific word or phrase linked to a SharePoint item. Each term has a unique ID, although it can have many associated labels or synonyms. There are two types of terms:
    - **Managed terms**: These are predefined or curated sets of terms created and managed by a Term Store administrator. Managed terms are selected from a list of terms to be applied to content.
    - **Enterprise Keywords**: These are words or phrases that a user adds to an item. Keywords can be used in a folksonomy or informal tagging mechanism.
- **Term set**: This is a group of related terms. Terms are typically organized in some sort of a hierarchy—for example, `Books > Cookbooks > Vegetarian`. Term sets can be open (anyone can add terms) or closed (only specific users can add terms). In addition, term sets can have two different scopes:
    - **Local**: Only available within the context of a specific site collection, such as with a specific list or library
    - **Global**: Available to all sites that use the Managed Metadata Service
- **Groups**: These are sets of term sets that share common security, business, or management requirements. To manage term sets in a specific group, a user must have contributor permissions for that group.
- **Tagging**: This is the process of applying a term (either from the Managed Metadata Service or Enterprise Keywords) to an item.
- **A Managed Metadata column**: A column is a data structure that holds data (such as terms or other metadata), similar to how you might add a column called `Notes` to a spreadsheet. A Managed Metadata column displays terms from a specific term set to use when tagging data. It can use an existing term set or be mapped to a new local term set specifically for that column.
- **An Enterprise Keywords column**: This is a type of column that can be added to content (such as libraries or lists) that allows users to tag items with their own words.

- **The Term Store management tool**: The Term Store management tool is used to manage term sets, terms, groups, and other aspects of the Managed Metadata Service. To manage the Term Store, a user must be a member of the Term Store administrators.
- **Content types**: A content type represents both a data item and information about that particular data item. Content types can be used to link or associate a particular item with related metadata, document templates, or policies.

Now that you have an understanding of the terminology for this section, you can learn how to configure a Managed Metadata Service application.

Planning a Managed Metadata Service application

Understanding your organization is paramount when defining the structure of the Managed Metadata Service. Managed metadata can be used to help enforce a certain consistency and structure across your organization, or it can be used in more grassroots style management to help users find resources by using terminology that is familiar to them. The core decision comes down to how you want to control the use of metadata in your organization:

- Do you want to control things globally or only use metadata locally on certain sites?
- Will term sets be curated and managed or will they be open for users to edit and contribute?
- Will you allow the use of Enterprise Keywords (folksonomy) or not?

SharePoint allows organizations to combine both formal, managed metadata with user-driven tagging, depending on the organizational data classification requirements.

Managed terms allow users to select from pre-defined terms, helping enforce consistent usage across sites (or the organization as a whole). It helps improve discoverability and search capabilities since you can be assured that the metadata values are spelled and applied consistently across the organization. Additionally, managed metadata allows metadata-based navigation, allowing users to filter search results on selectable options.

Planning for taxonomy and folksonomy

As mentioned previously, one of the benefits of managed term sets is the ability to apply metadata values to items consistently. Planning and implementing a term set helps users labeling content to use the same terms throughout the enterprise, eliminating incorrect or misspelled terms.

In creating a term set for your taxonomy, you'll need to understand how data is classified in your organization. To get an idea of how you might plan a taxonomy, let's look at this data table of cuts of beef that a restaurant supply company might use:

Section	Cut	Recommended Preparation
Chuck	Flat Iron	Grill
Chuck	Chuck Roast	Braise
Round	Rump Roast	Braise
Round	Sirloin Tip	Roast
Short Loin	Hangar Steak	Grill
Short Loin	T-Bone Steak	Grill
Short Plate	Short Ribs	Braise
Short Plate	Skirt Steak	Grill
Sirloin	Coulotte Steak	Sauté
Sirloin	Ball Tip Roast	Roast

Depending on how your organization refers to this data, one taxonomical structure may be preferred over another. For example, you may want users to navigate the terms in the previous term set by preparation method, followed by cut, and then section.

From a content classification perspective, you may want to enable users to tag certain cuts of beef with meal ideas (such as tagging both skirt steak and flat iron steak with the term `fajitas`). Once you have settled on a taxonomy structure and whether or not to allow Enterprise Keywords, you can begin configuring the Managed Metadata Service application.

Configuring the Managed Metadata Service application

In order to start using metadata on your farm, you'll need to configure the Managed Metadata Service. The steps in the following sections will walk you through a basic configuration if one hasn't already been created.

Creating and configuring a service account

First, you'll need to select (or create) an account for the purpose of running the Managed Metadata Service:

1. For this section, you can create a normal domain user account in **Active Directory Users and Computers**. We'll register it as a managed account assigned to the Managed Metadata Service. You can also use PowerShell to create an account:

   ```
   $Password =
   ([System.Web.Security.Membership]::GeneratePassword(15,2))
   $SecurePassword = ConvertTo-SecureString -AsPlainText $Password -
   Force
   New-ADUser -DisplayName "Managed Metadata Service Account" -
   SamAccountName mms-svc -Name "Managed Metadata Service Account" -
   AccountPassword $SecurePassword -Enabled $True
   ```

   You can retrieve the plain-text password in the $Password variable for use in *Step 6*. If you create the account through **Active Directory Users and Computers**, note down the password.

2. Launch **SharePoint Central Administration**.
3. Under **Application Management**, select **Manage service applications**.

4. Select **Security**, then select **Configure managed accounts** under the **General** section:

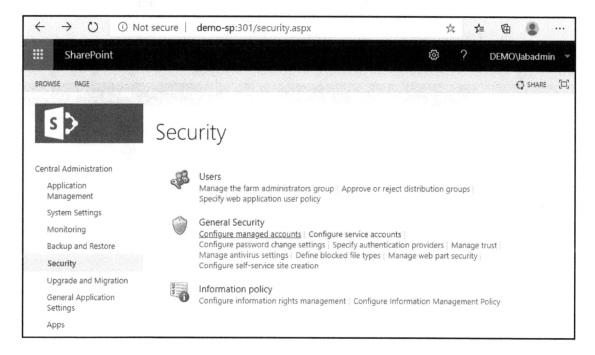

5. Select **Register Managed Account**.

6. Enter the credentials for the service account, using `DOMAIN\username` as the username and the password acquired from *Step 1*. If desired, select **Enable automatic password change** to configure SharePoint to automatically update the password on a scheduled basis, then scroll to the bottom of the page and click **OK**:

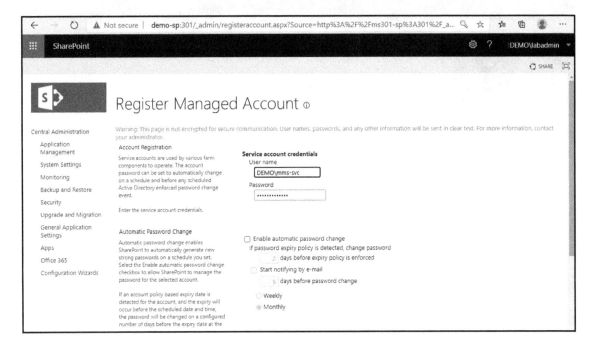

Now that you've created and configured an account, you can move on to creating the service application.

Configuring a Managed Metadata Service application

In this section, we'll instantiate the Managed Metadata Service application:

1. Click on **Application Management** from the left-side navigation menu.
2. Under **Service Applications**, select **Manage service applications**:

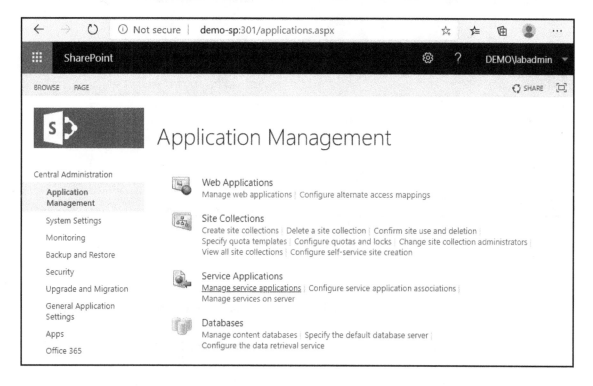

3. Click **New**, then select **Managed Metadata Service** from the service application list:

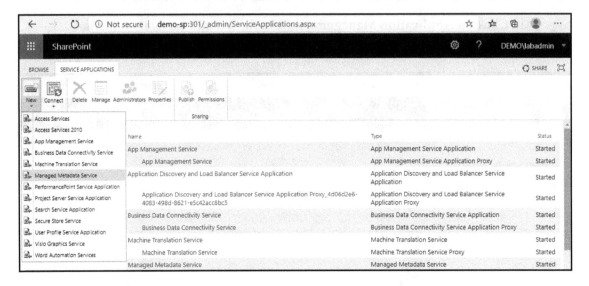

4. On the **Create New Managed Metadata Service** page, fill out the details as follows:

- **Name**: Enter a name for the service application.
- **Database Server**: Enter the instance of SQL Server where you want to create the managed metadata database.
- **Database Name**: Enter the name that you want to use for the managed metadata database.
- **Database Authentication**: Select either Windows or SQL authentication (supply credentials if you are using SQL authentication).
- **Failover Database Server**: Enter the name of your failover database server if you're using one.
- **Application Pool**: Select the **Create new application pool** radio button.
- **Application Pool Name**: Enter a name for the new application pool.
- **Select a security account for this application pool**: From the drop-down list, select the Managed Metadata Service account that you registered in *Step 6* of the previous section.
- **Content Type Hub**: If you're going to connect to a content type hub, enter the URL for that site collection in this box. If this is the first Managed Metadata Service application that you're configuring in a farm, leave this blank:

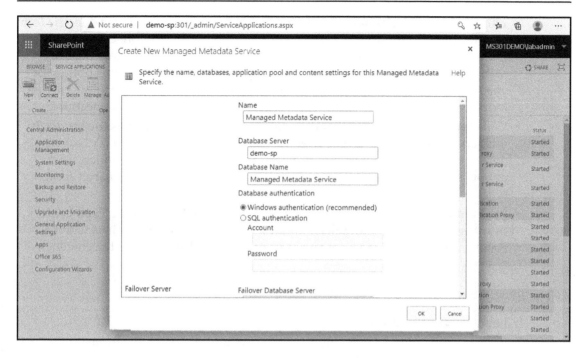

5. Click **OK**.

The Managed Metadata Service application has now been created. Next, we'll configure the settings for the Managed Metadata Service connection.

Configuring the Managed Metadata Service connection

After you have created one or more Managed Metadata Service applications, you can configure the settings for them. Configuring a service connection allows you to manage the following settings:

- **This service application is the default storage location for Keywords**: If you are using Enterprise Keywords (folksonomy) as part of your metadata strategy, you can enable this option to select this Managed Metadata Service application as the default storage location for Enterprise Keywords. If you've configured multiple Managed Metadata Service applications in your organization, you can only select this on one service application per web application. If you are *not* going to use Enterprise Keywords, you need to ensure this checkbox is cleared for all Managed Metadata Service connections.

- **This service application is the default storage location for column-specific term sets**: Enable this option if this service application is used to store custom term sets that are created at the site collection level. If you are using more than one Managed Metadata Service application in your farm, only select this option for one service application per web application. If you do not want to allow custom term sets, clear this checkbox for all Managed Metadata Service connections.
- **Consumes content types from the content type gallery at http://<sitename>**: If this service application will be used to make content types from a content type hub available to users of sites in this web application, select this option. This option is only available if the service has been configured to use shared content types.
- **Push-down Content Type Publishing updates from the Content Type Gallery to sub-sites and lists using the content type**: This option determines whether content type changes are published to sub-sites and lists that use the content type.

To configure the service connection, follow these steps:

1. Launch **Central Administration**.
2. Under **Application Management**, select **Manage service applications**.
3. Find the Managed Metadata Service connection you wish to configure and click on the row:

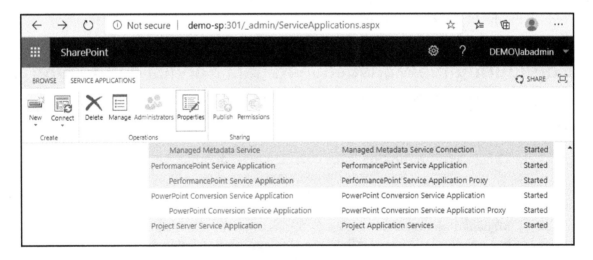

4. With the row highlighted, click on **Properties** under the top ribbon.
5. Select the checkboxes for the enable or disable options and then click **OK**:

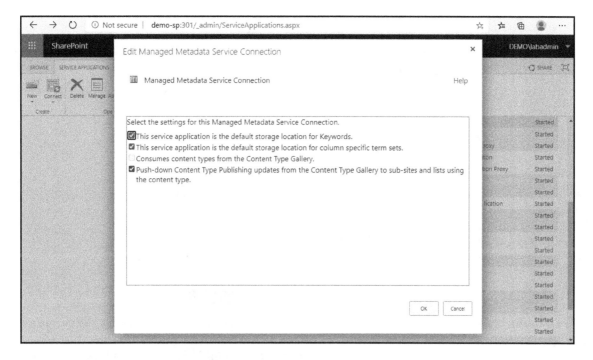

Now that you've configured a Managed Metadata Service application, you can begin configuring the Term Store.

Planning and configuring Term Store security

In the context of Term Store administration, a **group** is a collection of related terms. Every person who administers some aspect of the SharePoint Term Store must be granted some level of permission. By managing the Term Store security, you can assign or delegate administration and configuration of the entire Term Store, groups, and term sets. In this section, we'll configure the security of the Term Store, which is required for administering it later.

To make changes to the Term Store management tool, you must be granted one of the following specific roles:

- Term store administrator, which has the following capabilities:
    - Create or delete term set groups.
    - Add or remove group managers or contributors.
    - Change the working languages for the Term Store.
    - Carry out any other task that a group manager or contributor can do.
- Group manager, which has the following capabilities:
    - Add or remove contributors.
    - Perform any other task that a contributor can do.
- Contributor, which has the following capabilities:
    - Create or change a term set.

Let's begin with the Term Store administrator.

Adding a Term Store administrator

By default, SharePoint Server farms do not have any users in this (or any other Term Store administration) role. Before you can administer any Term Store features, you need to add an account to the Term Store administrator group. You should use an account that already has farm administration privileges to assign Term Store administrators. You need to be a Term Store administrator to assign other roles. Follow these steps to add Term Store administrators:

1. If this is your first time administering the Term Store, you'll need to grant account permission. In the tree-view navigation pane, select **Managed Metadata Service** (next to the home icon), add an account to **Term Store Administrators**, and click **Save**.

2. Navigate to the Term Store management tool. You can access it in a few ways. If you're already in **Central Administration**, you can go to **Application Management | Manage Service Applications** and then select the Managed Metadata Service application that you wish to administer. From a site collection, you can select **Site Settings | Term Store Management**:

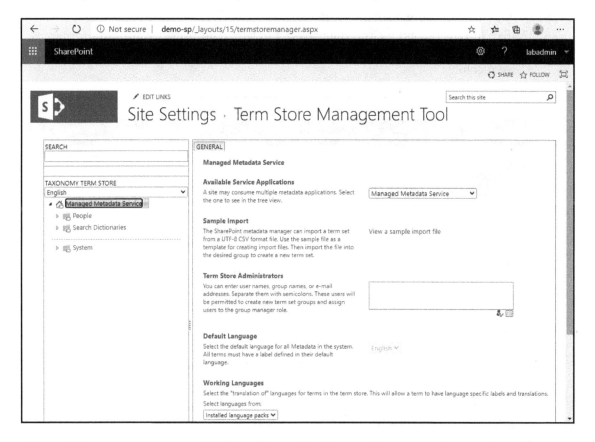

After adding your account to the Term Store administrators group, when you hover over items in the navigation, they'll now display a drop-down arrow. In the next section, we'll add group managers.

Adding a group manager

Group managers aren't necessary unless you are delegating roles to other users in the organization. Larger organizations may need to delegate the management of business-specific term sets to relevant groups. Follow these steps to add group managers:

1. Open **Term Store Management Tool**.
2. Expand the taxonomy and then select a group. Add users by adding usernames or addresses under the **Group Managers** section, separating them with semicolons:

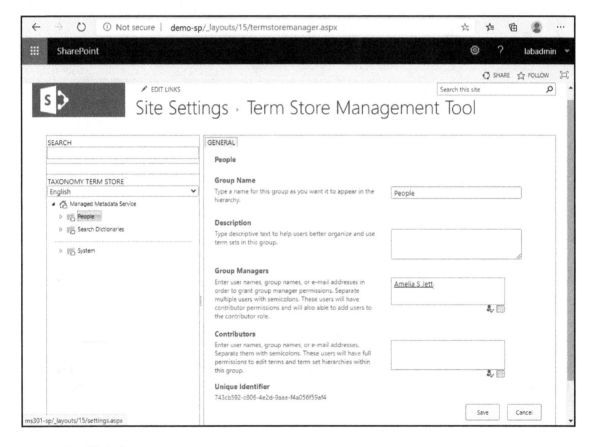

3. Click **Save**.

Let's see how to add contributors in the next section.

Adding a contributor

Similar to group managers, contributors aren't necessary for an organization (as either a group manager or Term Store administrator can contribute terms). Follow these steps to add contributors:

1. Open **Term Store Management Tool**.
2. Expand the taxonomy and then select a group. Add users by adding usernames or addresses under the **Contributors** section, separating them with semicolons:

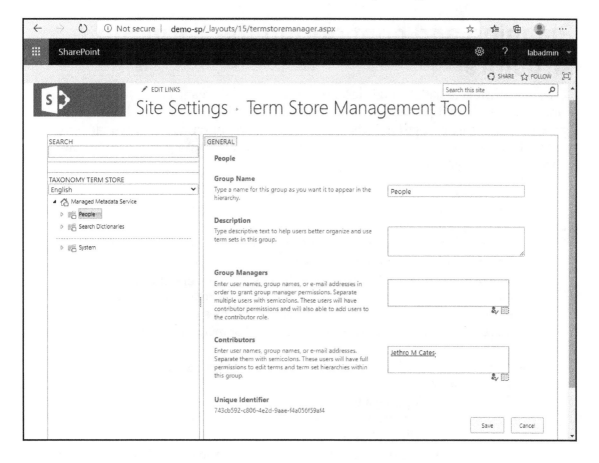

You can specify a person or a group as the contact, stakeholder, or owner of a term set. These labels do *not* grant any extra permissions to work with, edit, delegate, or otherwise manage the term set. They are only labels to identify business interests and stakeholders for the term set that may need to be consulted when making changes.

Once permissions have been assigned for managing the Term Store, you can begin defining the structure.

Planning and configuring the Term Store structure

While it is possible to have a single term set for your entire organization, you'll probably want to configure multiple term sets—perhaps each restricted to a certain business group, agency, or department. A group is the security boundary for term sets. When planning out the structure, use these recommendations and tips—they will give you a good balance between flexibility and structure:

- **Use groups**: While you may have global term sets (such as general business terms, department names, or other topics that are common across the enterprise), you may also find that departments or business groups have specific terms related only to their specific area of the organization—such as finance, legal, or human resources. You can use groups to organize term sets.

- **Use term set hierarchies**: Terms can be placed in a hierarchical fashion (such as `Books > Cookbooks > Italian` or `Construction > Materials > Roofing`). Complete and well-designed hierarchies will help users refine data more specifically and increase the likelihood of metadata being assigned correctly.

- Be careful to understand the difference between *copying, reusing,* and *pinning* terms:

  - **Copying** a term means you are using a source term as a template to create a target term. As soon as you save the new term, it is its own entity and is not connected to any other terms.

  - **Reusing** a term means you are creating an *editable linked copy* of it in another term set. If you later update the term (either the original or the copy), *both* terms get updated. This can be both useful and have unintended consequences.

- **Pinning** a term means you are making a *linked copy* of it, but the new term is read-only and can't be changed. This might be useful if you want to create standard sets of terms across your enterprise.

- Establish processes for handling the deprecation of old terms as your business changes.
- Use the **Other labels** property of a term to add synonyms to it, rather than creating the synonyms as new terms. This will help to reduce confusion and keep content labels consistent (for example, rather than creating separate terms for `lawyer`, `counsel`, and `attorney`, create one term for `lawyer` and then add the `counsel` and `attorney` synonyms to it).

Once you've determined how you want to manage the security and use (and reuse) of terms, you can begin creating term sets.

Creating and managing term sets

Before you can use term sets in your organization, you'll need to decide on a structure and architecture based on your business needs. The process to create and manage term sets is fairly straightforward. The regular maintenance procedures can be divided into the following categories:

- Creating a term set
- Managing a term set
- Creating a term
- Copying a term
- Reusing a term
- Pinning a term
- Merging a term
- Deprecating a term
- Moving a term or a term set
- Deleting a term or a term set

Let's go through each one of these individually.

Creating a term set

To create a term set, follow these steps:

1. Navigate to **Term Store Management Tool**. You can access it in a few ways. If you're already in **Central Administration**, you can go to **Application Management | Manage Service Applications** and then select the Managed Metadata Service that you wish to administer. From a site collection, you can go to **Site Settings | Term Store Management**:

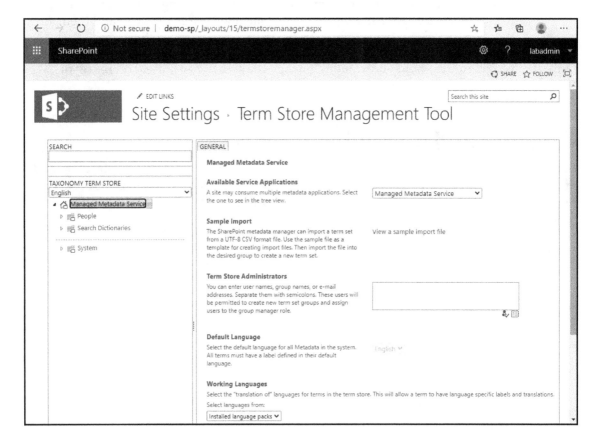

2. If this is your first time administering the Term Store, you'll need to grant account permission. In the tree-view navigation pane, select **Managed Metadata Service** (next to the home icon), add an account to **Term Store Administrators**, and click **Save**. After adding your account to **Term Store Administrators**, when you hover over items in the navigation, they'll now display a drop-down arrow.

3. Click on **Managed Metadata Service**, then select **New Group**:

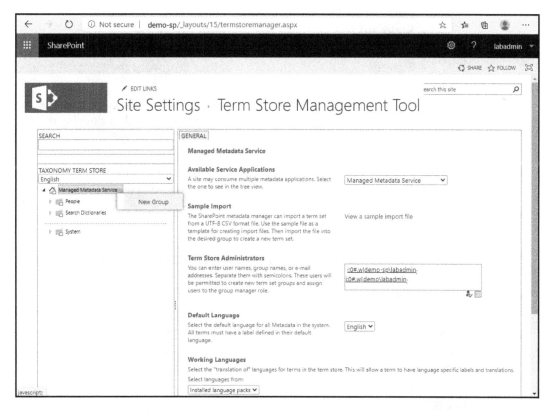

4. Fill out the properties for the name of the group managers and contributors and then click **Save.**

5. Point to the new term set (or an existing one if you have previously configured a group to be associated with a site), select the arrow that appears, and then select **New Term Set**.

6. Enter the value you want to use as the default label or name of your term.

7. In the **Properties** pane of the **Term Store Management Tool**, click on the **General** tab and then fill out the required information about the new term set:

   - **Term Set Name**: Enter a value for the name for your term set.
   - **Description**: Enter a description for using the terms.
   - **Owner**: Specify an owner.
   - **Contact**: Enter an email address of a user or group for feedback purposes.
   - **Stakeholders**: Add the names of groups or users that should be notified prior to changes being made to the term set.
   - **Submission Policy**: Specify whether you want the term set to be open or closed.
   - **Available for Tagging**: Select the checkbox to make the terms available for tagging items. If the term set is still in development or not ready to be deployed, clear the box.

8. Select the **Intended Use** tab, then specify the following settings:

   - **Available for Tagging**: Select this checkbox to make this term set available for tagging content.
   - **Use this Term Set for Site Navigation**: Select this checkbox to enable usage for managed navigation.
   - **Use this Term Set for Faceted Navigation**: Check this box to enable refiners based on managed properties on the managed navigation pages. This is an advanced setting.

9. Select the **Custom Sort** tab, then select a preferred sort order. By default, terms will be sorted alphabetically for the current language. Select **Use custom sort order** so that you can organize terms to always appear in a consistent order.

10. Select the **Custom Properties** tab to specify any additional shared property name and value data about the term set.

11. Click **Save**.

Managing a term set and maintaining the Term Store

There are a number of maintenance activities that you may need to perform throughout the life cycle of your SharePoint Term Store. These changes might be necessary as the organization implements changes to business units, departments, products, or services. Common tasks in a Term Store include creating, copying, reusing, pinning, merging, deprecating, moving, and deleting terms.

All of the following tasks are performed within the **Term Store Management Tool**.

Creating a term

Terms must be added to the Term Store before they can be applied as metadata to content. Use this process to add a new term to the Term Store:

1. Expand the groups to find the term set that you want to add a term to.
2. Point to the term set where you want to add a term, click on the down arrow that appears, and then select **Create Term**:

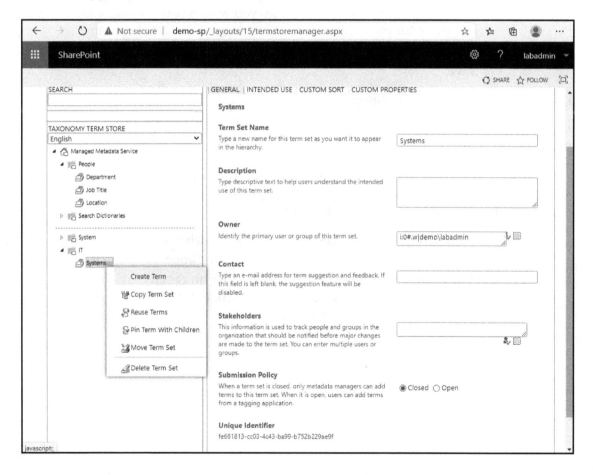

3. Enter the value that you want to use as the default label for the new term.
4. In the **Properties** pane, specify the following information about the new term:
   - **Available for tagging**: Select this box to make this term available for use in tagging.
   - **Language**: Select the language for this label. If you have not yet enabled multilingual terms, you'll only see the default language for your tenant.
   - **Description**: Enter a description of the term.
   - **Default Label**: Enter the default name of this term.
   - **Other Labels**: Enter any synonyms in the current language for this term. Synonyms are any values you want a search to associate as equal (for example, if you want `schoolroom` to be treated the same as `classroom`).
5. Click **Save**.

The new term is available for editing.

Copying a term

You may have terms that require similar properties. You can save time by copying the original term to use as a template. Use the following process to copy a term. The new term is placed in the same term set as the original term:

1. Expand the groups and navigate to the term you wish to copy.
2. Point to the term, then click on the down arrow that appears next to it.
3. Select **Copy Term**. The default label for the new term is `Copy of <original term>`. None of the term's child terms are copied, but the properties of the original term are copied to the new term:

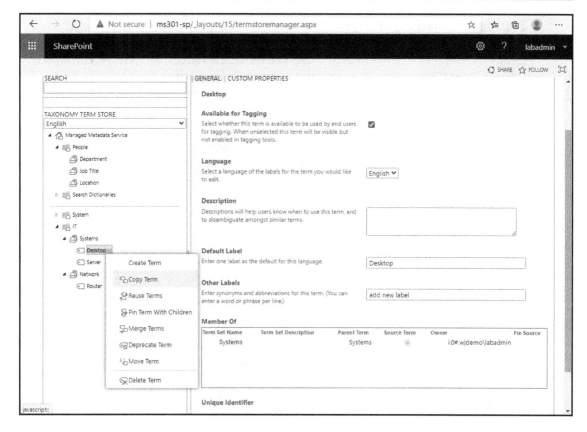

4. Make any additional changes and then click **Save**.

The new term should now be available for editing.

Reusing a term

As mentioned earlier, reusing a term basically involves creating a linked copy of a term in one term set to another term set. Take the following steps to reuse a term:

1. Expand the groups and navigate to the target term set where you want to reuse an existing term. The term you want *should not* be in this term set.

2. Point to the target parent item (either a term set or a parent term under which you want the new reused term to appear) and click on the down arrow that appears next to it, then select **Reuse Terms.** In the following example, the **General** term set under **Finance** is going to be the target for the **Standard Operating Procedures** term located in the **Marketing** term set:

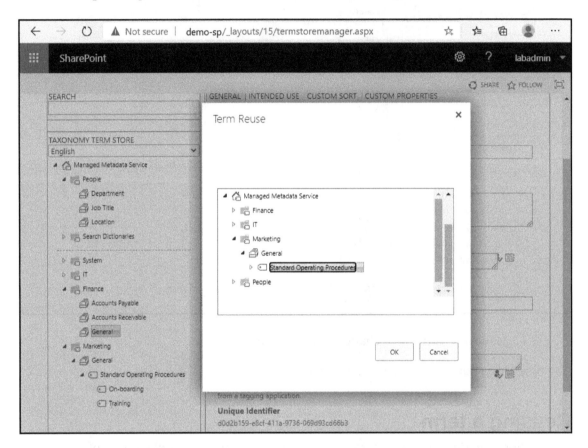

3. On the **Term Reuse** page, navigate to the source term. In this example, the source term is located in the **General** term set under the **Marketing** group. Select **OK**:

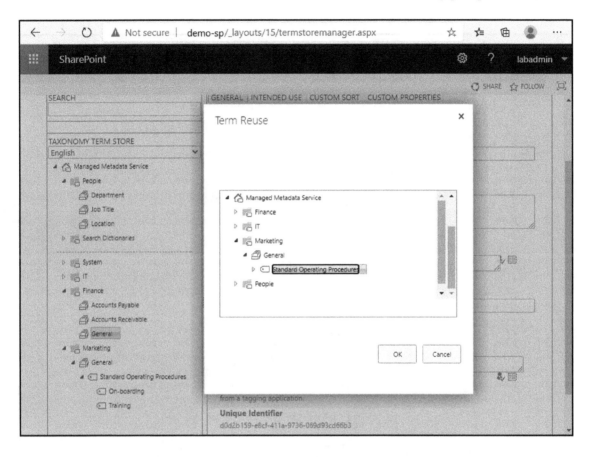

4. Notice the new term under the *target* term (or term set). The icon shows that it is a *linked* or *reused* term. Changes made to either of the linked terms will be reflected in the other linked term:

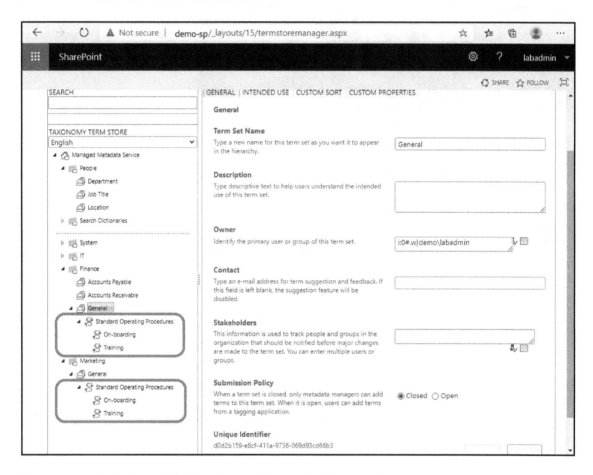

You can see both the original and reused terms in the console.

Pinning a term

Pinning a term is similar to reusing a term—the main exception being that only the parent (source) term can be modified. Take these steps to pin a term:

1. Expand the groups and navigate to the target term set where you want to pin an existing term. The term you want *should not* be in this term set.

2. Point to the target parent item (either a term set or a parent term under which you want the new pinned term to appear) and click on the down arrow that appears next to it, then select **Pin Term With Children**. In the following example, the **General** term set under **Marketing** will be the target for the **Managers** term located under the **Finance** term set:

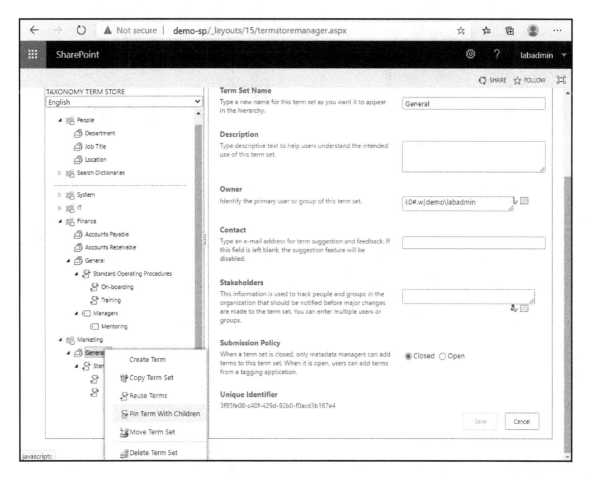

3. On the **Reuse and Pin Term: Select a Source** page, select the source term, then click **OK**:

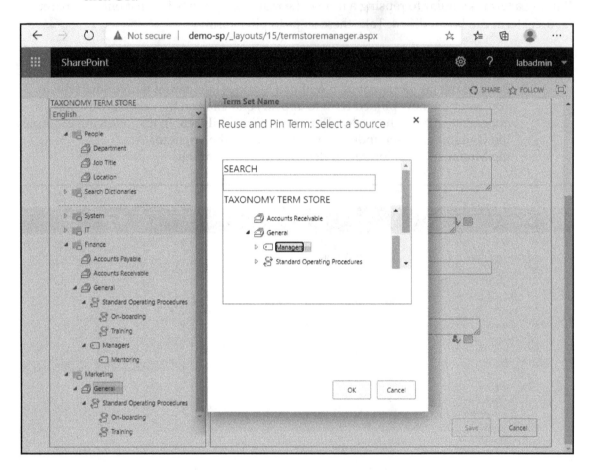

 Notice that the new term's properties are read-only.

Merging a term

If you have some level of redundancy in your terms, it may be time to merge them. You may need to merge terms if you are retiring one product and replacing it with a similar product or if you have duplicative terms that refer to the same concept. In either of these cases, merging terms could be the solution. To do so, take the following steps:

1. Expand the groups and navigate to the term set that contains the term you want to collapse into another.

2. Point to the item you wish to collapse or merge into another, click on the down arrow that appears next to it, and then select **Merge Terms**. In the following example, the **Time Off** term is going to be merged into the **Time Away** term:

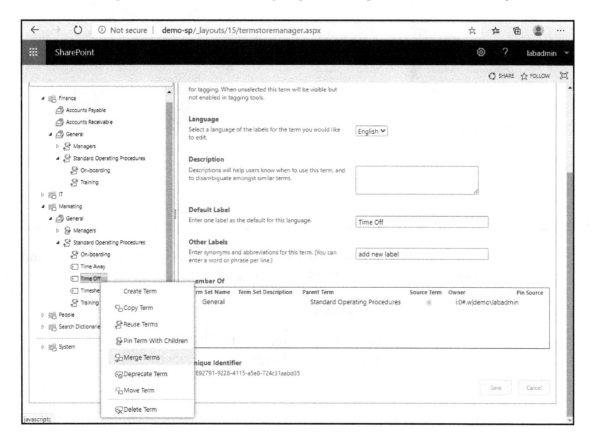

3. On the **Select Term to Merge Into** page, select the target term, then click **OK**:

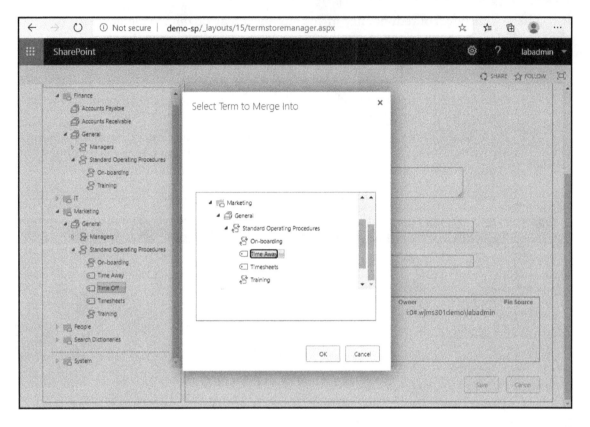

4. Notice the dialog box regarding the term merge. The synonyms, translations, and other properties of the source term will be copied into the target term. Click **OK** to continue the merge or **Cancel** to abort.

5. Click **Save** to save the updated term.

 Notice in the properties of the target term that the source term's label and synonyms have been added to the target term's properties.

Deprecating a term

At some point, it may become necessary to discontinue the future use of an existing term. Deprecating a term inactivates it for usage, but existing instances of the term continue to be applied. Deprecating a term does not affect any of its child terms:

1. Expand the groups and navigate to the target term set containing the term you want to deprecate.
2. Point to the item you wish to deprecate, click on the down arrow that appears next to it, and then select **Deprecate Term**. If you choose to deprecate a reused term, all instances of it are deprecated.

> You cannot deprecate a pinned term, though you can deprecate a reused term. You can also reactivate a deprecated term by selecting it and then choosing **Enable Term**.

Moving a term or term set

Take the following steps to move terms or term sets:

1. Expand the groups and navigate to the item you want to move.
2. Point to the item you wish to move, click the down arrow that appears next to it, and then select **Move Terms** (if the item is a term) or **Move Term Set** (if the item is a term set).
3. On the **Term move: select a destination** page, select the destination, then click **OK**.

> You can move terms between term sets or groups and can move term sets between groups.

Deleting a term or a term set

Deleting a term removes all instances of it. Items that had been previously tagged with the deleted term will not be tagged with that term anymore:

1. Expand the groups and navigate to the item you wish to delete.

2. Point to the item you wish to delete, click on the down arrow that appears next to it, then select **Delete Term** (if the item is a term) or **Delete Term Set** (if the item is a term set).

3. Click **OK** to acknowledge the deletion.

Let's see how term sets are imported in the next section.

Importing term sets

If you have already created a taxonomy elsewhere or have a very long term set to create, you can import it directly into the Term Store. The term set import file is a CSV file that has a specific format.

The first line of the CSV import contains the following comma-separated data:

```
"Term Set Name","Term Set Description","LCID","Available for Tagging","Term
Description","Level 1 Term","Level 2 Term","Level 3 Term","Level 4
Term","Level 5 Term","Level 6 Term","Level 7 Term"
```

The second line of the import file contains the actual name of the term set, a comma, a description of the term set, two more commas, either TRUE or FALSE to indicate whether the terms are available for tagging, and then eight more commas.

The third line of the import file is where the actual terms begin. The first two fields are blank, followed by the locale identifier (if this field is blank, then the locale of the Term Store is used by default), either TRUE or FALSE to indicate whether a term is available for use, and the term description. The last seven fields are used to show the hierarchy of the items.

Imagine the following bulleted list represents the items in the term set:

- Locations (term set)
    - United States (level 1 term)
        - New York (level 2 term)
            - New York City (level 3 term)
            - Albany (level 3 term)
            - Rochester (level 3 term)

- Florida (level 2 term)
    - Jacksonville (level 3 term)
    - Tampa (level 3 term)
    - Orlando (level 3 term)
- Europe (level 1 term)
    - France (level 2 term)
        - Nice (level 3 term)
        - Paris (level 3 term)
    - Germany (level 2 term)
        - Berlin (level 3 term)
        - Frankfurt (level 3 term)

The sample CSV import file might look as follows:

```
"Term Set Name","Term Set Description","LCID","Available for Tagging","Term
Description","Level 1 Term","Level 2 Term","Level 3 Term","Level 4
Term","Level 5 Term","Level 6 Term","Level 7 Term"
"Locations","Office Locations",,TRUE,,,,,,,,
,,1033,TRUE,,"United States","New York",,,,,
,,1033,TRUE,,"United States","New York","New York City",,,,
,,1033,TRUE,,"United States","New York","Albany",,,,
,,1033,TRUE,,"United States","New York","Rochester",,,,
,,1033,TRUE,,"United States","Florida",,,,,
,,1033,TRUE,,"United States","Florida","Jacksonville",,,,
,,1033,TRUE,,"United States","Florida","Tampa",,,,
,,1033,TRUE,,"United States","Florida","Orlando",,,,
,,1033,TRUE,,"Europe",,,,,,
,,1033,TRUE,,"Europe","France",,,,,
,,1033,TRUE,,"Europe","France","Nice",,,,
,,1033,TRUE,,"Europe","France","Paris",,,,
,,1033,TRUE,,"Europe","Germany",,,,,
,,1033,TRUE,,"Europe","Germany","Berlin",,,,
,,1033,TRUE,,"Europe","Germany","Frankfurt",,,,
```

You cannot add additional properties, such as synonyms, to the import file. The file must be saved in UTF-8 format. You can also download a sample of the import file by opening the Term Store management tool, selecting the taxonomy, and then selecting the **View a sample import file** link.

To import the term set file, follow these steps:

1. Open the Term Store management tool.
2. Navigate to the group where you wish to import the term set file. If one does not exist, you can create a group using the steps in the previous section. Point to the group, click on the down arrow that appears next to it, and select **Import Term Set**.
3. On the **Term Set Import** page, click **Browse** and navigate to the location of the saved import file. Click **OK**.
4. Review the imported term set:

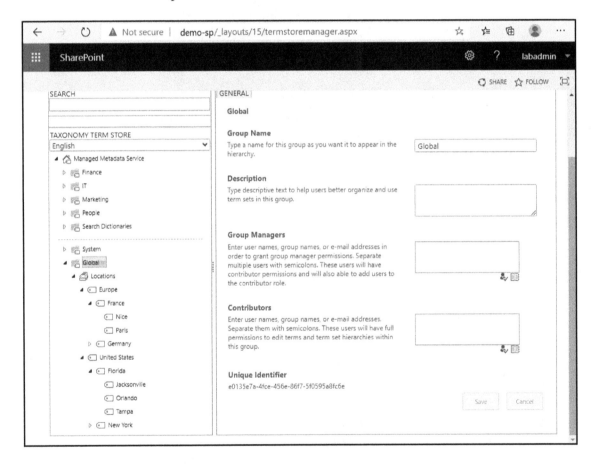

After importing a term set, you can go through and configure the properties of the terms as you would any other term in the store.

Next, we'll look at the tasks surrounding managing Term Store languages.

Planning and configuring Term Store languages

If your organization has locations internationally, chances are you'll need to support multiple languages in the Term Store. Multilingual support for terms means that users can tag and refer to content in their local languages, making it available to others who speak the same language. By default, the only language support available for SharePoint Server 2016 or SharePoint Server 2019 is the language that the installation was performed in.

The following sections will help you understand the process of deploying and managing languages in SharePoint.

Downloading and installing language packs

Adding support for additional languages in SharePoint Server requires downloading additional language packs. To do so, follow these steps on each server in your farm:

1. Download one or more additional language packs for SharePoint Server 2016 (`https://go.microsoft.com/fwlink/?LinkId=746633clcid=0x409`) or SharePoint Server 2019 (`https://www.microsoft.com/en-us/download/details.aspx?id=57463`). You'll want to create a separate folder for *each language pack you download*, as the language packs all share the same filename (`serverlanguagepack.exe`).

2. Run the language pack installation by double-clicking the `serverlanguagepack.exe` file. The installation will be displayed in the language of the language pack, so you'll need to be aware that the text will not match the installed server language. Click on the box at the lower left-hand side of the screen to accept the license terms and then click the button to continue:

3. Leave the box selected to run the SharePoint Products configuration wizard and then click on the button to finish:

4. Click **Next** on the **Welcome to SharePoint Products** page.
5. Click **Yes** to acknowledge that services will be restarted during configuration.
6. Click **Finish** to close the SharePoint Products configuration wizard.
7. Repeat the process for every additional language pack.

Additional languages *will not be available for use* until the SharePoint Products configuration wizard has been rerun.

Configuring multilingual support

Once you have installed additional language packs, you'll need to make them available on the SharePoint system for use. To configure multilingual support for your Term Store, follow these steps:

1. Open the Term Store management tool, select the top-level (taxonomy), and then under **Working languages**, select the additional languages you wish to support. If you have not installed any additional languages, none will show up under **Installed language packs**:

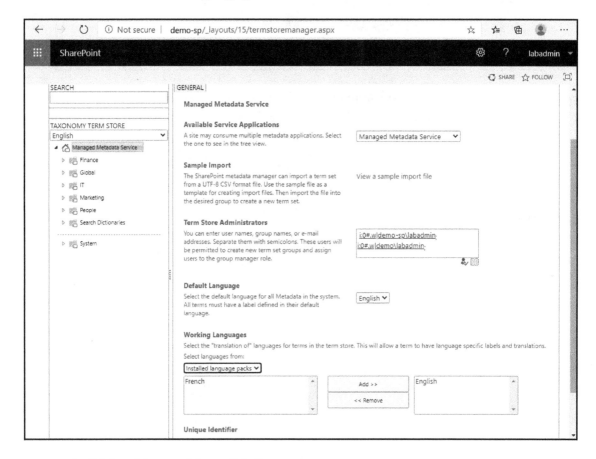

2. Click **Add >>** and then click **Save**.

Once you have enabled multilingual support, you can then begin adding labels in other languages to your terms. In the following example, we will add a French label for the Resources term:

1. Select the term in the navigation tree.
2. From the **General** tab, under **Language**, select the additional language that you wish to add labels or synonyms for:

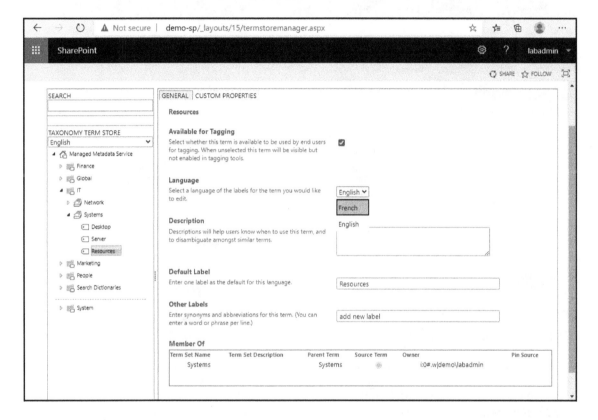

3. Every language requires a default label. Enter the primary label for the newly selected language. In this case, the French variation of **Resources** is **Ressources**. Add any additional synonyms or labels as you would for the primary language. When finished, click **Save**:

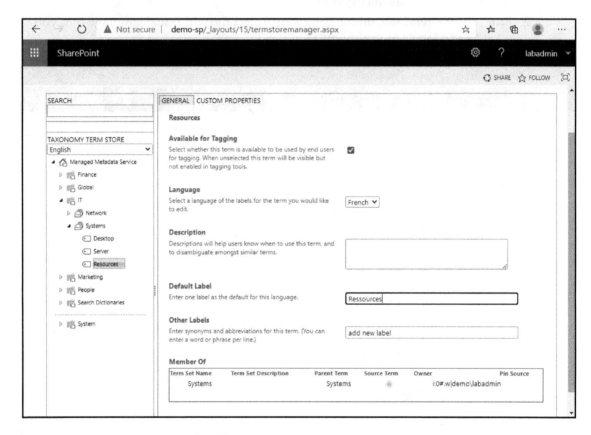

You can continue adding additional synonyms to other terms throughout the store, thereby improving the discoverability, searchability, and accessibility of content throughout your organization.

Summary

Managed metadata can be used to provide a consistent database of structured terms to apply to content throughout an organization, improving search and discovery experiences. In this chapter, you learned about the capabilities of the Managed Metadata Service, as well as learning about common administrative tasks, such as creating, copying, and deleting terms. In addition, you learned about the multilingual capabilities of the Managed Metadata Service, further improving accessibility and search capabilities for those whose native language is different than that of the content creator.

In the next chapter, we'll learn about how to configure the SharePoint Online environment for guest access and some of the methods available to secure it.

8
Managing Search

In this chapter, we're going to explore the role of the search service in SharePoint Server. As organizations experience more data proliferation (documents, spreadsheets, images, databases, and other sorts of content), keeping a list of bookmarks or memorizing URLs becomes a daunting task. The role of search in the SharePoint environment is to help users locate the resources they need quickly and efficiently.

SharePoint search works by crawling lists and libraries, indexing content, and adding site columns and values to the search index. Once inside the search index, the site columns are mapped to managed properties. When users perform searches, the query is sent to the search index, which then returns the results to the search results page.

SharePoint Search has two different search experiences – one for classic SharePoint and one for modern SharePoint. Modern Search creates a more personalized search, based on the content you have access to and your previous Office 365 activity. Classic search, on the other hand, uses the search results web part, while modern search does not. As you saw in Chapter 3, *Managing and Maintaining a SharePoint Farm,* you perform customization differently between the modern and classic experiences.

When choosing which search experience to promote for your users, consider these factors:

- Modern search doesn't support filters or formatting for organization-specific content types or company name extraction.
- Classic search supports custom refiners and search verticals.
- Modern search lets users search across connected hub sites.
- Modern search doesn't support changing the sort order of results or metadata-based refiners.

The structure, organization, and design of your SharePoint environment will help drive which search experience will work best for your users.

In this chapter, we're going to explore the following topics:

- Creating and updating search dictionaries
- Managing query suggestions
- Managing result sources
- Managing the search schema
- Managing the search center settings
- Monitoring search Usage Reports and Crawl Logs

Let's get started!

Overview of Search

Search is the service application responsible for indexing content and making it available to users via a discovery mechanism, such as a search center. Search has several components and sub-processes, including the following:

- **Crawl**: The crawling process is responsible for moving through connected data sources to gather content for processing.
- **Analytics**: Analytics processing covers two sub-components: usage and search. Analytics can be used to help refine the relevance of content for users, as well as to provide metrics for reporting.
- **Content processing**: The content processing component handles document parsing, as well as linguistics processing sub-components for language detection and entity extraction. Content processing then hands data over to the indexing component.
- **Index**: The search index is the storage component that contains extracted entities and data. The index components receive the output of content processing and return results to the queries.
- **Query processing**: The primary function of the query processing component is to analyze a query and then submit it to the index component. The query processing component of the search also handles linguistics processing tasks, such as word breaking and stemming.
- **Administration**: This is responsible for the system-level processes related to search, such as provisioning search instances and running the actual search process.

Next, we'll start looking at the components of search and some of its functions, starting with the search schema.

Managing the search schema

In SharePoint Server, the search schema determines how content is ingested from the search index. In `Chapter 7`, *Planning and Configuring Managed Metadata*, we discussed how metadata is *data about data*. That metadata is extracted during the crawling process, including both structured content (derived from properties of documents, pages, or other data) and unstructured content (extracted keywords). The search schema defines the configuration of what users can search for and how the results are presented. SharePoint search returns only what is stored in the search index. The results are security-trimmed, meaning they're only what the user has access to see.

The SharePoint search index, such as your favorite internet search tool's index, is populated by crawling sites and the content of your SharePoint Online sites. The content and metadata make up the properties of an item, while the schema's *crawled properties* instruct the search crawler on what to retrieve and store.

Crawled properties are mapped to *managed properties*, and those managed properties are what the search index contains. By default, the SharePoint search service has many managed properties defined. However, if you want to add additional managed properties, you'll need to map them to crawled properties so that the data can successfully flow into the search index and be discoverable.

Administrators have the ability to create new custom-managed properties, but the content is limited to either text or Yes/No values. Other content types are available, but you'll need to reuse one of the existing built-in (but unused) managed properties. You can rename the unused managed properties with an alias.

There are a number of settings for managed properties (including *searchable, queryable, retrievable, sortable, and refinable*). Not all of the managed property settings apply to both classic and modern experiences, so if you are going to start configuring managed properties, you should refer to the table at `https://docs.microsoft.com/en-us/sharepoint/search/search-schema-overview` for more information.

You can manage the search schema for either a site collection or the entire farm. The overall schema will apply to both classic and modern search. In this example, we'll create a simple managed property for *writer* and map it to the crawled property *author*. Let's get started:

1. Navigate to the **Search Schema** page for a site collection or the tenant:
    - For a site collection, navigate to the site collection, click the gear icon, select **Site information**, select **Site Settings**, and select **View all site settings**. Locate **Search Schema** and select it.

- For the farm, navigate to **Central Administration** | **Application Management** | **Manage service applications** | **Search Service Application** and click on **Search Schema**.

2. Click **New Managed Property**.

3. Enter a value in the **Property name** box (since we're mapping *writer* to *author*, you can name the property Writer):

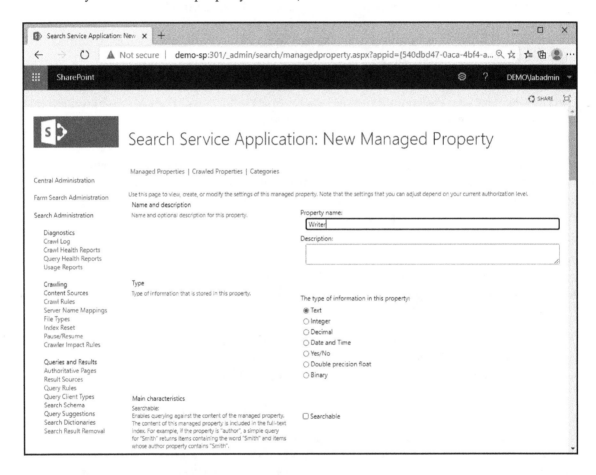

4. Under **Type**, select **Text**.
5. Under **Main characteristics**, select the checkboxes for **Searchable**, **Queryable**, and **Retrievable**.
6. Under **Mappings to crawled properties**, click **Add a mapping**:

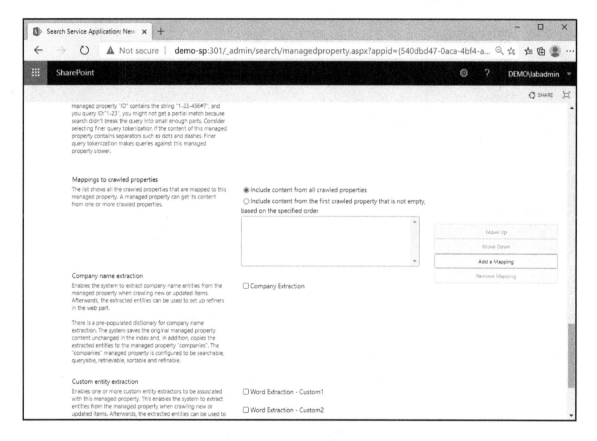

7. On the **Crawled property selection** page, locate the **Author** property and click the **Next** arrow at the bottom of the list of crawled properties:

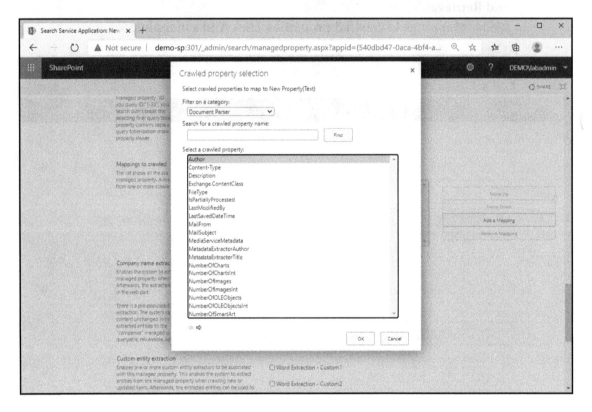

8. Click **OK** to close the **Crawled property selection** page.
9. Click **OK** to save.

Now, you'll be able to search for content using `writer:<name>` as a way to filter content.

Next, we'll review the search dictionaries.

Creating and updating search dictionaries

Search dictionaries are used to manage content that is extracted from documents. Company name extraction and query spellings are two examples of how search dictionaries are used.

Company name extraction is a feature that pulls company name entities from content and maps them to the **companies** managed property. Customizing this feature only affects classic search.

Query spelling suggestions allow you to configure words that you wish to include or exclude for spelling corrections. Query spelling suggestions also only work for classic search.

The following examples demonstrate some of the scenarios where search dictionaries can be used.

Managing company names

Company name extraction works by finding company names and mapping them to the *companies* managed property. For example, if a company name is found in the body of a document, the company name is extracted. You can then use the *companies* managed property to create refiners using the company name in the **Refinement Web Part** on a classic search results page.

SharePoint Server comes with a pre-populated dictionary for company name extraction. You can add additional company names to be extracted by placing their names in the **Company Inclusions** section of **Term Store**. You can also prevent the extraction and mapping of a company name by placing it in the **Company Exclusions** list.

In order to manage company name extraction, you'll need to configure the company names in **Term Store**. To complete this task, you will need to be a Term Store administrator (see Chapter 7, *Planning and Configuring Managed Metadata*, for more information). To update the company names, perform these steps:

1. Launch **Central Administration**, select **Application Management**, select **Manage service applications**, and then select **Managed Metadata Service** to open **Term Store Management Tool**:

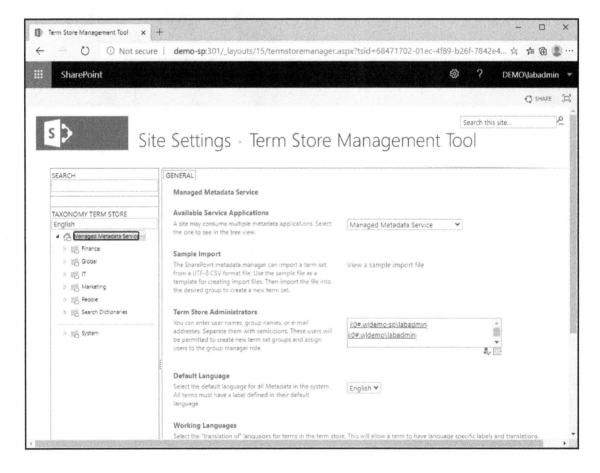

2. Expand **Search Dictionaries** in the navigation pane.
3. Point to either **Company Inclusions** or **Company Exclusions**. Then, select the down arrow that appears and click **Create Term**:

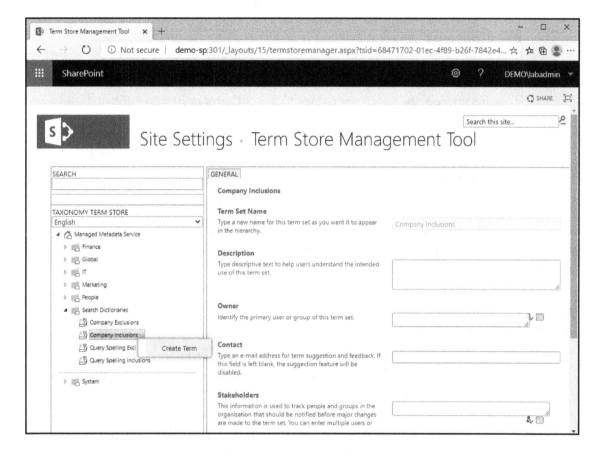

4. Enter the name of the company to include or exclude, and either click outside the typing area or press *Enter*.

 If you are considering migrating to Microsoft 365 or migrating content to SharePoint Online, as of November 2019, company extraction has been deprecated. See `https://docs.microsoft.com/en-us/sharepoint/` `changes-to-company-name-extraction-in-sharepoint-online` for more information.

As indexing is performed, the company names listed in **Company Inclusions** or **Company Exclusions** will be processed accordingly.

Managing query spelling

Query spelling, also known as "Did you mean?", is a tool that can help users locate resources, even when they've misspelled what they're looking for. SharePoint Online has a dynamic query dictionary, but it only identifies words that appear in at least 50 documents by default. To include (or exclude) query spellings, you'll need to use the Term Store Management tool's Search Dictionaries feature.

Under either area, you'll enter the *spellings of terms you want to have suggested* (under **Query Spelling Inclusions**) or the *spellings of terms you never want to have suggested* (under **Query Spelling Exclusions**). To do this, perform the following steps:

1. Launch **Central Administration**, select **Application Management**, select **Manage service applications**, and then select **Managed Metadata Service** to open **Term Store Management Tool**.
2. Expand **Search Dictionaries** in **Term Store Management Tool**.
3. Point to either **Query Spelling Inclusions** or **Query Spelling Exclusions**. Then, select the down arrow that appears and click **Create Term**:

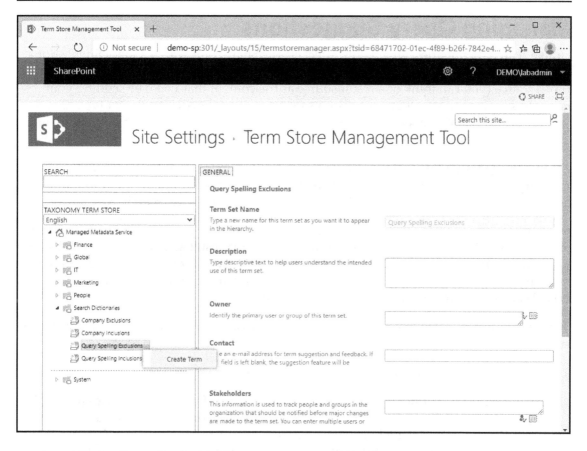

4. Enter the spelling of the query word to include or exclude, and then click outside the typing area or press *Enter*.

The updated dictionary settings should be available in about 10 minutes and should be reflected in user searches.

In the next section, we'll work on managing search further with query suggestions.

Managing Query Suggestions

Unlike search dictionaries, which only affect classic SharePoint search, Query Suggestions customization affects both classic and modern Search, since they can use the same default result source (you can think of a result source as the content that you'll be sending your query to search against). Query suggestions (also known as predictive text or type-ahead words) appear below the search box as a user enters their query words. SharePoint Online creates a query result for a suggested term when you've clicked a result at least six times.

The query suggestions are populated daily per result source and per site collection, meaning that the suggestions can be different across the organization. These automatically generated suggestions for the default result source are used in both search experiences.

In addition to the automatically generated query suggestions, you can also create your own lists and upload them. These manual lists apply to all the result sources and site collections for both search experiences. If you want to create suggestions for multiple languages, you'll need to create a separate query suggestion file for each language. The files must be saved in UTF-8 encoding format to be successfully imported.

Similar to search dictionaries, you can create lists of words and phrases that will always be suggested, as well as lists of words and phrases that will never be suggested.

To manage search query suggestions, perform the following steps:

1. Using a text editor such as Notepad, create a list of terms and phrases to either *always suggest* or *never suggest*. Enter each term or phrase on its own line. When you're finished, save the text file in UTF-8 format.
2. Launch **Central Administration**, select **Application Management**, select **Manage service applications**, and then click **Search Service Application**.
3. Under **Queries and Results**, select **Query Suggestions**:

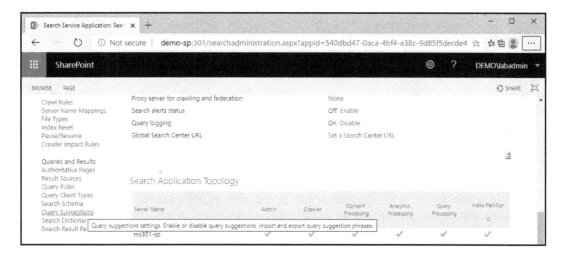

4. On the **Search** page, under **Always suggest phrases** and **Never suggest phrases**, select **Import from text file**:

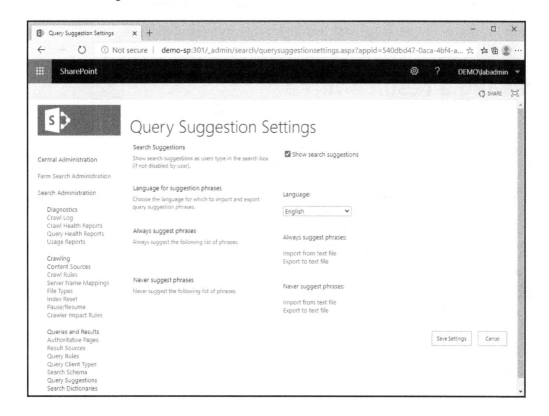

5. Click **Browse** and select the file you created previously. Then, click **OK.**
6. Click **OK**.
7. Finally, click **Save Settings**.

The query suggestions will be updated and should be visible to your users shortly. Next, we'll look at using result sources to deliver more scoped content results.

Managing result sources

Result sources can be used to limit the scope of searches to a certain set of content or to a particular subset of search results. Result sources may also be configured to include external search services and to send queries to third parties. Both classic and modern Search experiences can use the default result source, but you can only create result sources for the classic search experience. If you change the default result source for the tenant, it will affect both the classic and modern Search experiences. You can use a feature called query transform to restrict result sources to a particular subset of content as well.

You can configure a result source for the farm, a site collection, or a site.

Creating a result source

To configure a result source, perform the following steps:

1. Locate the **Manage Results Source** page for the scope of the search:
   - For the farm, this is located in **Central Administration (Application Management | Manage service applications| Search Service Application | Result Sources)**.
   - For either a site collection or a site, navigate to the site, click the gear icon, select **Site information**, select **Site Settings**, and select **View all site settings**. Locate either **Search Result Sources** (site collection) or **Result Sources** (site).
2. Click **New Result Source.**
3. Fill out a name and description for the result source.

4. Select a protocol – most likely, you'll be configuring **Local SharePoint** (the default), but you do have other options. If you want to configure another SharePoint farm as a result source, select **Remote SharePoint**. You can also select **Exchange server index** (and you'll be prompted to supply an EWS URL). Finally, you can choose OpenSearch 1.0/1.1 to send queries to an external search service, such as Bing, that supports the OpenSearch protocol. In this example, we're going to select **OpenSearch 1.0/1.1** as the protocol so that we can send queries to the Bing search engine:

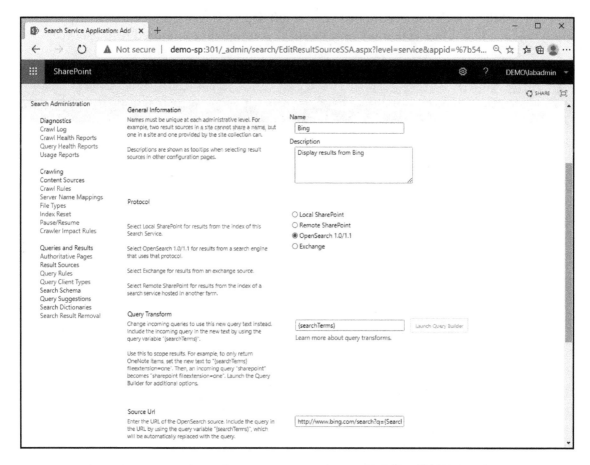

5. The default query transform is `{searchTerms}`, but you can use the query builder to create your own. If you launch the Query Builder tool, you can specify query filters on the **BASICS** tab, sort options on the **SORTING** tab, and then test out the query on the **TEST** tab.

6. Depending on the protocol selected, you may need to populate additional fields, such as **Remote Service URL (Remote SharePoint), Source URL (OpenSearch), Exchange Source URL (Exchange), or authentication**. In this example, we used the **Source URL** value of `http://www.bing.com/search?={searchTerms}+site:packtpub.com&format=rss&Market=en-US`. The effect is sending whatever search terms to `bing.com` and only returning results from the site packtpub.com.

7. Click **Save.**

You can test the search and ensure that your result source is configured correctly by querying for content that you know only exists in your newly configured result source:

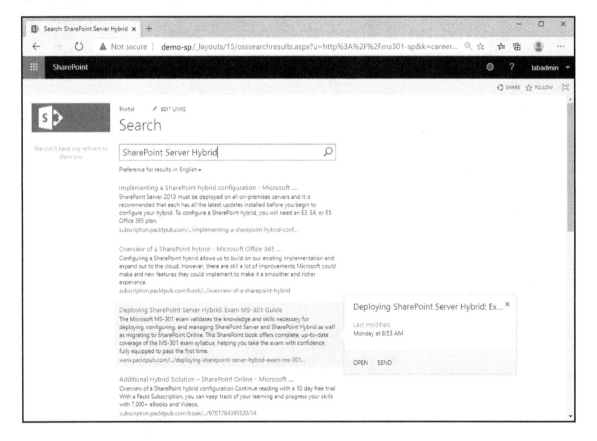

After creating a custom result source, you can create custom query rules that will use this result source.

Managing query rules

A query rule is a set of conditions and actions that help meet the user's search intent. Query rules allow you to promote or highlight certain types of results based on keywords, phrases, or content types. You can define query rules for the tenant, a site collection, or a site, and they can be applied to one or more result sources.

In the following example, we're going to create a search query to send specific terms to the result source we created in the previous section.

To create a query rule, perform the following steps:

1. Navigate to the **Manage Query Rules** page for a site, site collection, or the tenant:
   - For either a site collection or a site, navigate to the site, click the gear icon, select **Site information**, select **Site Settings**, and then select **View all site settings**. Locate either **Search Query Rules** (site collection) or **Query Rules** (site).
   - For the farm, this is located in **Central Administration** (**Application Management** | **Manage service applications** | **Search Service Application** | **Query Rules**).

2. Select a result source for the new query rule from the dropdown labeled **Select a Result Source**:

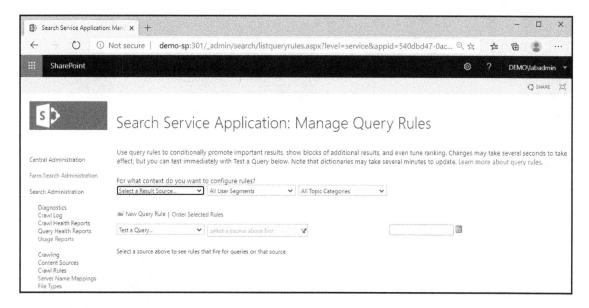

3. Click **New Query Rule**.

4. On the **Add Query Rule** page, enter a name for the query rule.

5. Click to expand the **Context** area.

6. In the **Context** section, select a result source. By default, the result source that you chose in the dropdown earlier is selected, but you can add more result sources or modify the ones that are already present. You can change it to all result sources by selecting **All sources**, or you can add additional result sources by clicking **Add Source**.

7. In the **Context** section, you can also choose to restrict the query rule to a category by selecting **Add category**.

8. In the **Context** section, you can restrict the query rule to a user segment by selecting **Add User Segment**.

9. In the **Query Conditions** section, configure a condition:

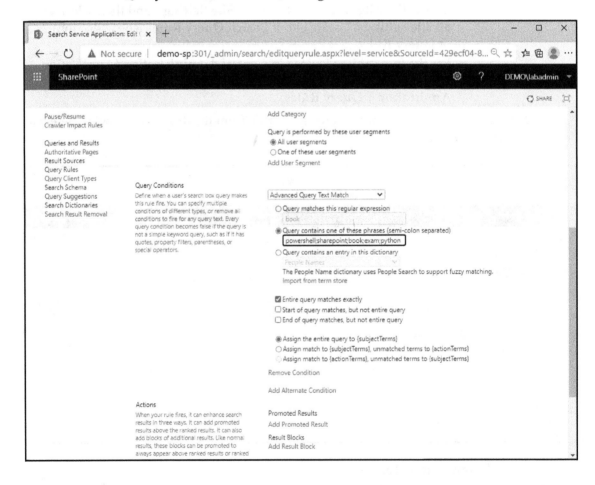

10. In the **Actions** area, specify the action to take. In this case, we're going to promote a few results (in previous versions of SharePoint, this was called *Best Bets*). Promoted results are links to content designed to drive traffic to specific sites based on the administrator's goals. Click **Add Promoted Result** and specify a name for the result in the **Title** field and then a URL. Then, click **Save**.

11. Under the **Actions** area, you can also add additional result blocks to use other result sources:

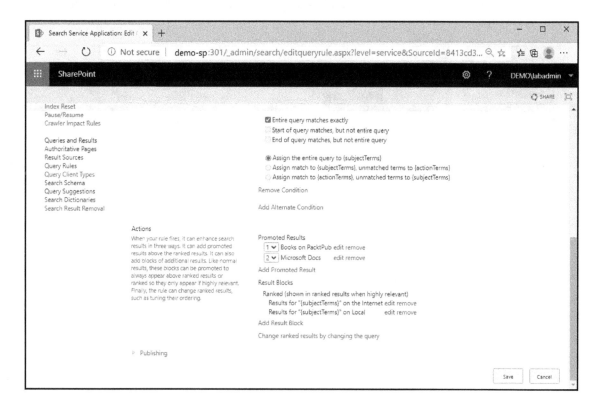

12. If you want to apply the rule for a limited amount of time, expand **Publishing** and select start and stop dates.
13. Click **Save**.

The next time a user conducts a search that matches your query rule, the promoted result will appear at the top. In this case, we created a promoted result based on the `powershell` keyword. To see the promoted result, you can go to a site and perform a search. If you use the keywords specified in the query rule, you'll generate a hit, as shown in the following screenshot:

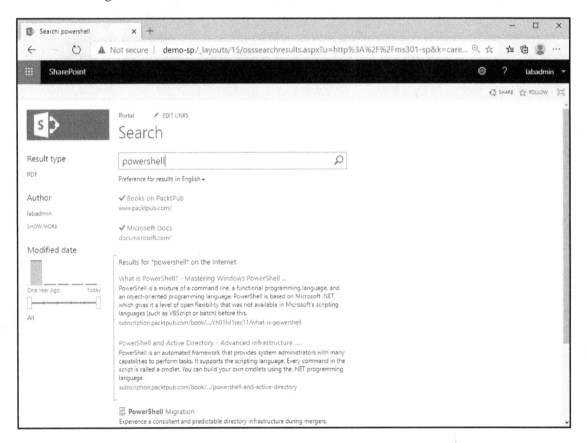

Now that you've learned how to manage some of the result source capabilities, we can start looking at configuring the result output.

Configuring Search and result output

In this section, we'll review some of the configuration possibilities for the Search web part, as well as the result types and display templates. These options will help you set the query and display parameters for searches submitted on the SharePoint farm.

You can customize the search experience further by modifying the properties of the various web parts that are used to input search terms and display the results, including the following:

- The Content Search web part
- The Search Results web part
- The Search Results query
- The Search Results display template
- The Search Results settings
- The Search Results refinement

Let's examine each of these in turn.

The Content Search web part

The Search Box is the web part that you normally think of for searching: it's an input textbox like you'd see on an internet search engine. The Search Box web part is displayed on the default home page, as well as the default results pages. You can update the settings for the search box's output display, query suggestions, enable or disable links for preferences or advanced search, or the actual template that's used to display the web part. In the following examples, we're going to customize an enterprise search site. To create one, create a new site collection using the Enterprise Search template.

To modify the Search Box web part, perform the following steps:

1. While logged in as a user with admin rights, navigate to an existing search center page (if you don't have a search center, you can add the web part to any page), click the gear icon to expand the **Settings** menu, and then select **Edit Page**.

2. Select the **Search Box** web part, select the down arrow that appears, and then click **Edit Web Part**:

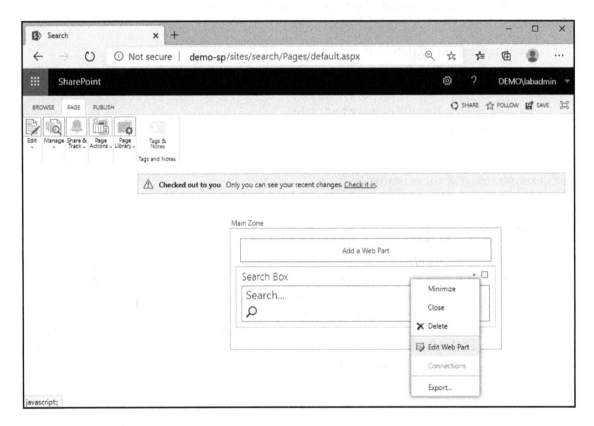

3. Under **Properties for Search Box**, you can configure the following:
   - Query destination, which will either be the current page, a custom URL, or a navigation node
   - Enable or disable query suggestions for content and names
   - Additional settings for displaying preferences, advanced search links, and page templates:

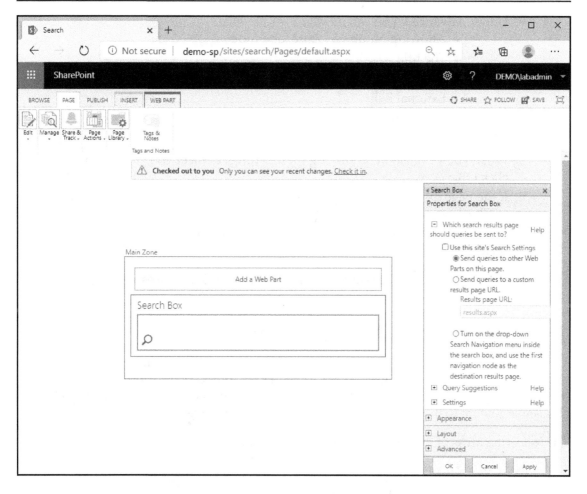

4. Under **Appearance**, you can configure the following:
   - The name of the search box
   - Fixed height and width
   - Box chrome (border) styling

5. Under **Layout**, you can configure the following:
   - Whether the box is hidden
   - Text directionality

6. Under **Advanced**, you can modify the following:
   - Allowing minimize, close, and hide
   - The title
   - The description
   - External help
   - URL link
   - Audience scoping

7. When you've finished modifying these settings to meet your requirements, click **OK**.

8. If the page displays a message stating that it is **Checked out**, click the link to **Check it in**.

9. If the page displays a message stating that it needs to be published, select **Publish**.

Now that you've configured the search box, let's look at the **Search Results** page.

The Search Results web part

The Search Results web part is used to display the output of searches in the classic search. It provides the formatting that's necessary for the preconfigured search verticals (**People**, **Everything**, **Videos**, and **Conversations**). In addition, the Search Results web part also sends content to the Refinement and Search Navigation web parts.

The following areas of the Search Results web part can be configured:

- Search Results query
- Search Results display template
- Search Results web part settings

We'll briefly examine all three areas. For all of the modifications, you'll need to edit the Search Results web part. Let's get started:

1. Navigate to the **Search Results** page. You can do this by launching classic search and then entering any search criteria. The default results page will be displayed.
2. Select the gear or **Settings** icon and select **Edit page**:

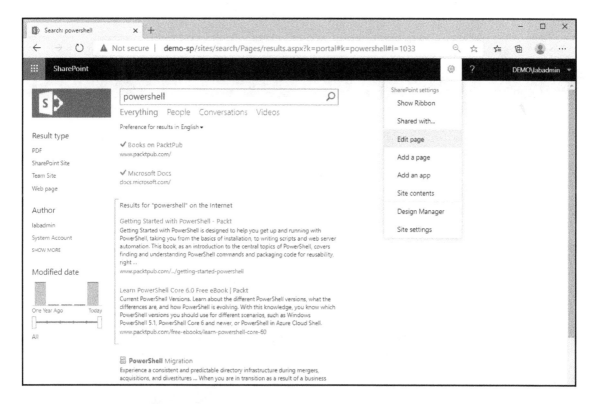

3. Select the **Search Results** web part, click the drop-down arrow, and then select **Edit Web Part**.

You can use this as the starting point for the next several configuration areas.

The Search Results query

The Search Results query is a piece of configuration that handles how the Search Results web part processes the search. The token variable, `{searchboxquery}`, passes the input from the Search Box web part to the Search Results web part. On the **Properties** page of the Search Results web part (right-click **Search Results** | **Edit Web Part**), you can modify the settings that are available, as shown in the following screenshot:

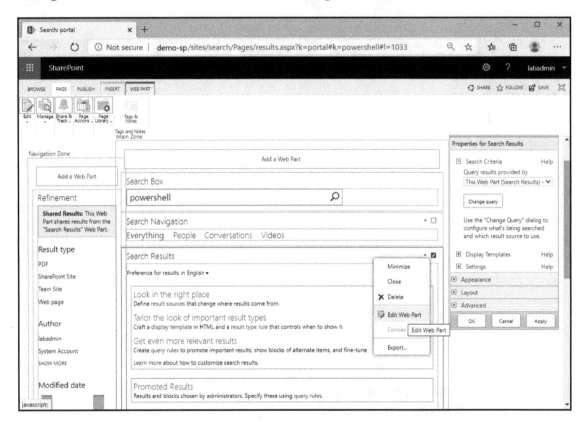

The **Change query** button under **Search Criteria** allows you to add additional query parameters, including result sources to use for processing queries in this search. Inside this dialog box, you can structure advanced queries using the **Kutso Query Language** (**KQL**) or quick queries using simple radio buttons and drop-down selectors. You can also add *refiners* (properties that are used for grouping, sorting, or categorizing content) and configure settings to remove duplicates, rewrite URLs, and whether or not you want to use the site's query rules.

The Search Results display template

By configuring display templates, you are controlling the display formatting of items. The main options are to let the result type determine the display options or to use a single template to format the display. Display templates show data from the managed properties of the result set, such as the following:

- File type
- Author
- Title
- Path
- Document summary
- Preview image

Display templates are divided into three categories:

- **Control display templates**: This template contains the structure of the results, such as where headings go, numbering, and navigation buttons.
- **Item display templates**: The item display template contains the managed properties whose data will be presented.
- **Hover panel display templates**: As the name indicates, the hover panel display template contains configuration information for the preview that's displayed when the mouse hovers over the result.

You can configure result types for a collection or a site by navigating to the settings for the site or collection and then selecting **Result Types** under the search area. Once you're in the **Result Types** area, you can configure display types for particular data types, as shown in the following screenshot:

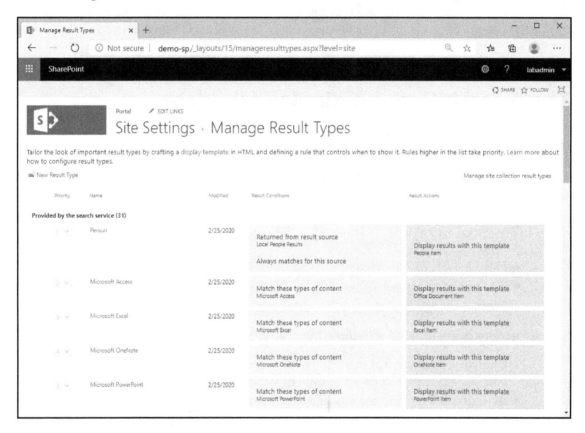

For more information on configuring display templates beyond the realms of this book, please refer to `https://docs.microsoft.com/en-us/sharepoint/use-result-types-and-display-templates`.

The Search Results settings

The **Settings** sub-option allows you to configure the number of results that are returned per page, whether to display links to the advanced search, the result counts, whether to display the sort option, whether the user's OneDrive site is included in the result's sources, and whether you want to display ranked or promoted results, as shown in the following screenshot:

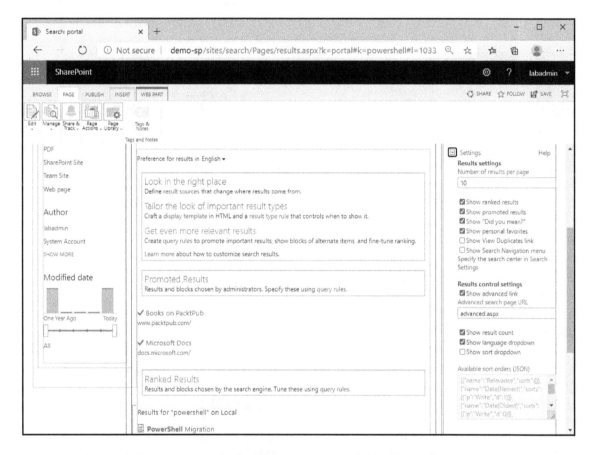

For more detailed descriptions of all of the options available for the Search Results web part, see https://support.office.com/en-us/article/change-settings-for-the-search-results-web-part-40ff85b3-bc5e-4230-b1dd-f088188e487e.

Search Results refinement

The Refinement web part is used to configure the refining, categorization, sorting, and grouping options for search results. You can use either a managed navigation term set or configure a list of refiners with the **Choose refiners...** button. The available refiners can be seen in the following screenshot:

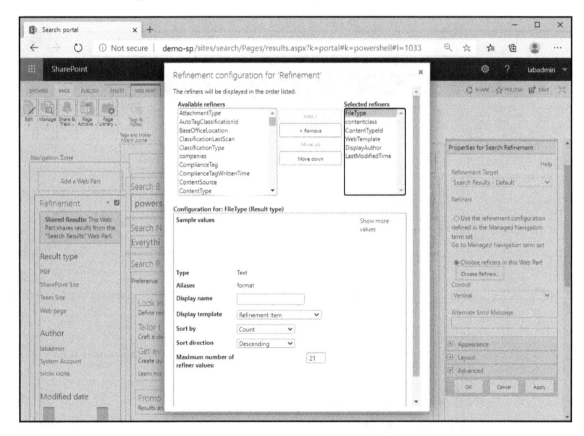

Search refinement will enable users to filter content based on selectable values, giving them more control over the final results that are displayed.

As shown in the previous sections, there are a number of differences regarding the capabilities of modern and classic search. Microsoft's direction is to move toward modern Search, but there are still gaps in how the experiences are delivered. As an administrator or developer, you may wish to explore the **SharePoint Framework (SPFx)** as a mechanism to bridge that gap. You can learn more about the SPFx at `https://docs.microsoft.com/en-us/sharepoint/dev/spfx/sharepoint-framework-overview`.

We'll learn how to configure search farms in the next section.

Configuring search farms

For organizations that rely on managing content through localized or departmentalized SharePoint farms or organizations that have millions of documents, it may be necessary (or desirable) to configure and deploy farms that are only responsible for handling search functions. In `Chapter 2`, *Planning a SharePoint Farm*, and `Chapter 3`, *Managing and Maintaining a SharePoint Farm*, we discussed basic MinRole farm architectures. In some enterprise scenarios, though, it may be necessary to dedicate multiple servers to processing and servicing search requests for your organization.

For these scenarios, you can start with the pre-built scenarios that Microsoft provides:

Volume of content (SharePoint 2013)	Volume of content (SharePoint 2016)	Sample Search architecture
Up to 10 million items	Up to 20 million items	Small search farm
Up to 40 million items	Up to 80 million items	Medium search farm
Up to 100 million items	Up to 200 million items	Large search farm
Not supported	Up to 500 million items	Extra-large search farm

Microsoft's reference architectures are designed to be deployed as **virtual machines (VMs)**, with two VMs per host, with the exception of database servers (either on physical hardware or 1:1 virtual machine:physical host). The following table lists the per-role requirements for each farm:

Small Search Farm				
Server Role	Hosts	Storage	Memory	CPU
Application server that has query processing and index components	A, B	500 GB	32 GB	1.8 GHz 8x cores
Application server that has crawl, search administration, analytics, and content processing components	A, B	200 GB	8 GB	1.8 GHz 4x cores

Database server that has all search databases (physical or single VM on a host)	C, D	100 GB	16 GB	1.8 GHz 4x cores
Medium Search Farm				
Server Role	**Hosts**	**Storage**	**Memory**	**CPU**
Application server that has query processing and index components	A, B, C, D	500 GB	32 GB	1.8 GHz 8x cores
Application server that has an index component	A, B, C, D	500 GB	32 GB	1.8 GHz 8x cores
Application server that has analytics and content processing components	E, F	300 GB	8 GB	1.8 GHz 4x cores
Application server that has crawl, search administration, and content processing components	E, F	100 GB	8 GB	1.8 GHz 4x cores
Database server that has all search databases (physical or single VM on a host)	G, H	400 GB	16 GB	1.8 GHz 4x cores
Large-Search Farm				
Server Role	**Hosts**	**Storage**	**Memory**	**CPU**
Application server that has query processing and index components	A, B, C, D, E, G, H	500 GB	32 GB	1.8 GHz 8x cores
Application server that has an index component	A, B, C, D, E, F, G, H, I, J	500 GB	32 GB	1.8 GHz 8x cores
Application server that has analytics and content processing components	K, L, M, N	300 GB	8 GB	1.8 GHz 4x cores
Application server that has crawl, search administration, and content processing components	K, L	100 GB	8 GB	1.8 GHz 4x cores
Database server that has search databases (physical or a single VM on a host)	O, P, Q, R	400 GB	16 GB	1.8 GHz 4x cores
Extra Large Search Farm				
Server Role	**Hosts**	**Storage**	**Memory**	**CPU**
Application server that has an index component	A-X	500 GB	32 GB	1.8 GHz 8x cores
Application server that has query processing and index components	Y, Z	500 GB	32 GB	1.8 GHz 8x cores
Application server that has crawl, search administration, or content processing components	AA-AF	100 GB	8 GB	1.8 GHz 4x cores
Application server that has analytics processing components	AG, AH	300 GB	8 GB	1.8 GHz 4x cores
Database server that has search databases (physical or single VM on a host)	AI-AL	400 GB	16 GB	1.8 GHz 4x cores

After installing and configuring the search farm with the SharePoint Products and Configuration Wizard, you'll need to go through a couple of additional steps:

- Configuring a Trust for Search
- Configuring a Remote SharePoint results source

We'll look at those steps briefly here.

Configuring a Trust for Search

In this section, we'll look at the security and authentication configurations necessary to enable the enterprise search farm (the *Receiving Farm*) to be able to receive, process, and return the results of queries from the content farm (the *Sending Farm*).

To configure the *Receiving Farm* to trust the *Sending Farm*, perform the following steps:

1. On a server in the *Receiving Farm*, launch **SharePoint Management Shell**.
2. Run the following script in the SharePoint Management Shell from any server in the *Receiving Farm*, updating the -Name and -MetadataEndPoint parameters with an appropriately chosen name in the -Name parameter and the address of a web app in the *Sending Farm*:

```
$i = New-SPTrustedSecurityTokenIssuer -Name "SendingFarm" -
IsTrustBroker:$false -MetadataEndpoint
"https://<SendingFarm_web_application>/_layouts/15/metadata/json/1"
New-SPTrustedRootAuthority -Name "SendingFarm" -MetadataEndPoint
https://<SendingFarm_web_application>/_layouts/15/metadata/json/1/r
ootcertificate
$realm = $i.NameId.Split("@")
```

3. Then, for each web application in the receiving farm that hosts content you want to search for, run the following command, replacing the value for <ReceivingFarm_web_application>:

```
$s1 = Get-SPSite -Identity https://<ReceivingFarm_web_application>
$sc1 = Get-SPServiceContext -Site $s1
Set-SPAuthenticationRealm -ServiceContext $sc1 -Realm $realm[1]
$p = Get-SPAppPrincipal -Site
https://<ReceivingFarm_web_application> -NameIdentifier $i.NameId
```

At this point, the authentication has been configured to allow the *Sending Farm* to be able to access the *Receiving Farm*.

Configuring a Remote SharePoint results source

In this configuration process, you'll use the steps in the *Managing result sources* section to create a Remote SharePoint results source on the *Sending Farm* that uses *Remote SharePoint* as the protocol, specifying the enterprise search farm as the remote SharePoint URL.

For more information on configuring a results source, refer to the *Managing result sources* section earlier in this chapter.

Next, we will walk through modifying the topology that Search uses.

Managing the search topology

Every SharePoint farm starts with a single server. If you are expanding your SharePoint infrastructure using one of the design patterns (such as one of the Search farm architectures) or your own schematic, you'll need to configure which components go where.

Managing the search topology is how you do that. In these next few sections, we'll review how to update or change the search topology based on your goals.

Prerequisites

Before we begin, you'll need to ensure that the following prerequisite configurations have been made:

- SharePoint Server is already installed on all of the servers that you want to use as part of your topology.
- You've mapped out which services will be installed on which hosts.
- Ensure that no crawls have been started and that the index has no content in it (you can verify this through **Application Management** | **Manage Service Applications** | **Search Service Application** | **System Administration** | **System Status**; **Searchable Items** should display 0).

For the purposes of configuring an example, we'll look at the Small Search Farm, as described in the *Configuring search farms* section:

Search Server A ("ServerA")	Search Server B ("ServerB")	Search Server C ("ServerC")	Search Server D ("ServerD")
Admin component 1	Query processing component 1	Admin component 2	Query processing component 2
Crawl component 1	Index component 1	Crawl component 2	Index component 2
Content processing component 1		Content processing component 2	
Analytics processing component 1		Analytics processing component 2	

Once we've done this, we can start configuring.

Configuring an enterprise search service instance

To start the configuration, you'll need to specify the servers you wish to start and configure the service on. Perform the following steps:

1. On a server in the farm you wish to configure, launch an instance of the SharePoint Management Shell.
2. Identify the servers (in this instance, we're using the server names from the table in the previous section) that will be used and configure a variable with their identities:

```
$hostA = Get-SPEnterpriseSearchServiceInstance -Identity "ServerA"
$hostB = Get-SPEnterpriseSearchServiceInstance -Identity "ServerB"
$hostC = Get-SPEnterpriseSearchServiceInstance -Identity "ServerC"
$hostD = Get-SPEnterpriseSearchServiceInstance -Identity "ServerD"
```

3. Start the search service on those servers:

```
Start-SPEnterpriseSearchServiceInstance -Identity $hostA
Start-SPEnterpriseSearchServiceInstance -Identity $hostB
Start-SPEnterpriseSearchServiceInstance -Identity $hostC
Start-SPEnterpriseSearchServiceInstance -Identity $hostD
```

4. Create a new search topology:

```
$ssa = Get-SPEnterpriseSearchServiceApplication
$newTopology = New-SPEnterpriseSearchTopology -SearchApplication
$ssa
```

5. Add the new components to the topology:

```
New-SPEnterpriseSearchAdminComponent -SearchTopology $newTopology -
SearchServiceInstance $ServerA;
New-SPEnterpriseSearchCrawlComponent -SearchTopology $newTopology -
SearchServiceInstance $ServerA;
New-SPEnterpriseSearchContentProcessingComponent -SearchTopology
$newTopology -SearchServiceInstance $ServerA;
New-SPEnterpriseSearchAnalyticsProcessingComponent -SearchTopology
$newTopology -SearchServiceInstance $ServerA;
New-SPEnterpriseSearchQueryProcessingComponent -SearchTopology
$newTopology -SearchServiceInstance $ServerB;
New-SPEnterpriseSearchIndexComponent -SearchTopology $newTopology -
SearchServiceInstance $ServerB -IndexPartition 0;
New-SPEnterpriseSearchAdminComponent -SearchTopology $newTopology -
SearchServiceInstance $ServerC;
New-SPEnterpriseSearchCrawlComponent -SearchTopology $newTopology -
SearchServiceInstance $ServerC;
New-SPEnterpriseSearchContentProcessingComponent -SearchTopology
$newTopology -SearchServiceInstance $ServerC;
New-SPEnterpriseSearchAnalyticsProcessingComponent -SearchTopology
$newTopology -SearchServiceInstance $ServerC;
New-SPEnterpriseSearchQueryProcessingComponent -SearchTopology
$newTopology -SearchServiceInstance $ServerD;
New-SPEnterpriseSearchIndexComponent -SearchTopology $newTopology -
SearchServiceInstance $ServerD -IndexPartition 0
```

6. Activate the newly created topology with the following command:

```
Set-SPEnterpriseSearchTopology -Identity $newTopology
```

7. Review the output of `Get-SPEnterpriseSearchTopology -SearchApplication $ssa -Text` to ensure that the components are active.

The search topology has been created and made active. You can also perform other activities, such as viewing or cloning a search topology, which we'll look at next.

Performing other search topology updates

If you need to change the configuration of your search farm, unfortunately, you cannot do so to a directly running topology. You must first *clone* your topology, make any updates to the clone, and then activate the clone. Here are some examples of activities.

Viewing the current active topology

You can view the makeup of the current active topology as follows:

1. Launch an instance of **SharePoint Management Shell**.
2. Run the following commands:

```
$ssa = Get-SPEnterpriseSearchServiceApplication
$active = Get-SPEnterpriseSearchTopology -SearchApplication $ssa -Active
Get-SPEnterpriseSearchComponent -SearchTopology $active
```

3. Finally, review the active components:

```
Administrator: SharePoint 2019 Management Shell                    —    □    ×

PS C:\> Get-SPEnterpriseSearchComponent -SearchTopology $active

IndexPartitionOrdinal  : 0
RootDirectory          :
ComponentId            : aab0e465-97ae-4fb7-9ba4-181dd3f6ac17
TopologyId             : d9d4b961-002d-47bd-8c04-fc83e2695974
ServerI                : afc3e3f4-0bc5-49a7-8935-e55ab43cad3b
Name                   : IndexComponent1
ServerName             : demo-sp
ExperimentalComponent  : False


NumProcessingTasksInParallel : 4
MaxSortMemoryInMB            : 400
ComponentId                  : af78bda4-a6e0-4b1e-9f82-1964cf5ab030
TopologyId                   : d9d4b961-002d-47bd-8c04-fc83e2695974
ServerId                     : afc3e3f4-0bc5-49a7-8935-e55ab43cad3b
Name                         : AnalyticsProcessingComponent1
ServerName                   : demo-sp
ExperimentalComponent        : False


ComponentId            : 2a0b9c12-ade1-495d-aaee-2ba4221a8266
TopologyId             : d9d4b961-002d-47bd-8c04-fc83e2695974
ServerId               : afc3e3f4-0bc5-49a7-8935-e55ab43cad3b
Name                   : CrawlComponent0
ServerName             : demo-sp
ExperimentalComponent  : False


ComponentId            : a3eef4de-8fa8-4a3c-bef2-40a70fecc60b
TopologyId             : d9d4b961-002d-47bd-8c04-fc83e2695974
ServerId               : afc3e3f4-0bc5-49a7-8935-e55ab43cad3b
Name                   : AdminComponent1
ServerName             : demo-sp
ExperimentalComponent  : False


ComponentId            : 621c43d2-e4b7-46d0-9e3e-b69951bd3c22
TopologyId             : d9d4b961-002d-47bd-8c04-fc83e2695974
ServerId               : afc3e3f4-0bc5-49a7-8935-e55ab43cad3b
Name                   : ContentProcessingComponent1
ServerName             : demo-sp
ExperimentalComponent  : False


ComponentId            : 31ac7966-4825-49da-b6f2-e11003ccd02d
TopologyId             : d9d4b961-002d-47bd-8c04-fc83e2695974
ServerId               : afc3e3f4-0bc5-49a7-8935-e55ab43cad3b
Name                   : QueryProcessingComponent1
ServerName             : demo-sp
ExperimentalComponent  : False
```

The output will display a list of components associated with the topology, including which servers each component is active on.

Cloning

Cloning the active search topology captures a copy of the current topology in a variable that you can make modifications to.

To clone the active search topology, perform the following steps:

1. Launch an instance of **SharePoint Management Shell**.
2. Run the following commands:

```
$ssa = Get-SPEnterpriseSearchServiceApplication
$active = Get-SPEnterpriseSearchTopology -SearchApplication $ssa -
Active
$clone = New-SPEnterpriseSearchTopology -SearchApplication $ssa -
Clone -SearchTopology $active
```

The cloned topology, stored in `$clone`, can be modified and activated.

Adding a component

Using the syntax in the *Configuring an enterprise search service instance* section, you can add additional components to servers in your cloned topology. You'll need either a new or existing *Search Service Instance* deployed to be able to add a component.

In this example, we'll start a new search instance on an existing SharePoint Server using `Get-SPEnterpriseSearchServiceInstance` and `Start-SPEnterpriseSearchServiceInstance`:

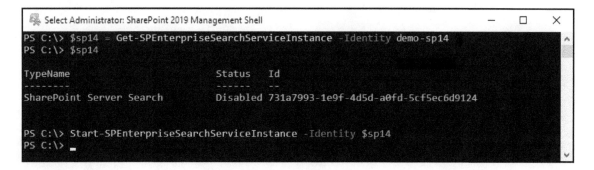

Once you have a search instance you wish to add components to, perform the following steps:

1. Identify which search components to enable. In this example, we'll add a Crawl component. You could also add other components such as Analytics, Query Processing, Search Admin, or Content Processing.
2. With the results of `Get-SPEnterpriseSearchServiceInstance` for the target server saved in a variable, run the following command (where `$<host>` is the variable containing the new search service instance):

```
New-SPEnterpriseSearchCrawlComponent -SearchTopology $clone -
SearchServiceInstance $<host>
```

The component will be added to the cloned configuration. To complete the update, you'll need to activate the topology.

Removing a component

To successfully remove a component from a cloned search topology, you'll need to perform the following steps:

1. Capture and clone the current topology using the following commands:

```
$ssa = Get-SPEnterpriseSearchServiceApplication
$active = Get-SPEnterpriseSearchTopology -SearchApplication $ssa -
Active
$clone = New-SPEnterpriseSearchTopology -SearchApplication $ssa -
Clone -SearchTopology $active
```

2. Review the components in `$clone`, identifying the clone you want to remove by passing the value to the `Get-SPEnterpriseSearchServiceComponent` cmdlet:

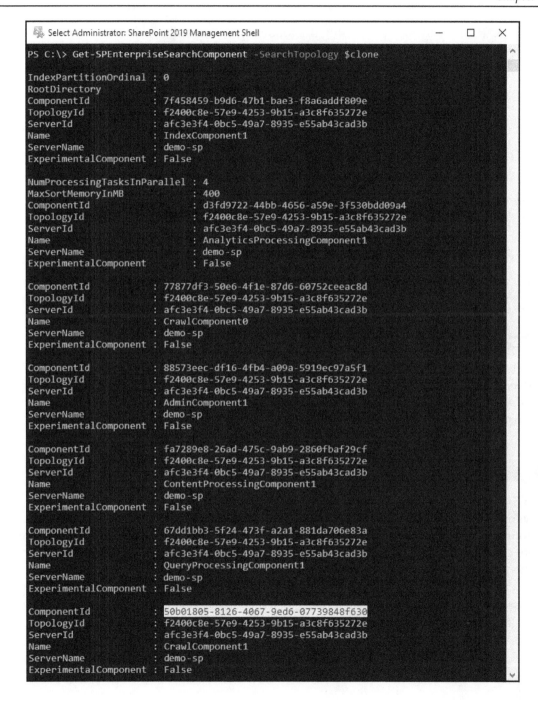

```
PS C:\> Get-SPEnterpriseSearchComponent -SearchTopology $clone

IndexPartitionOrdinal : 0
RootDirectory         :
ComponentId           : 7f458459-b9d6-47b1-bae3-f8a6addf809e
TopologyId            : f2400c8e-57e9-4253-9b15-a3c8f635272e
ServerId              : afc3e3f4-0bc5-49a7-8935-e55ab43cad3b
Name                  : IndexComponent1
ServerName            : demo-sp
ExperimentalComponent : False

NumProcessingTasksInParallel : 4
MaxSortMemoryInMB            : 400
ComponentId                  : d3fd9722-44bb-4656-a59e-3f530bdd09a4
TopologyId                   : f2400c8e-57e9-4253-9b15-a3c8f635272e
ServerId                     : afc3e3f4-0bc5-49a7-8935-e55ab43cad3b
Name                         : AnalyticsProcessingComponent1
ServerName                   : demo-sp
ExperimentalComponent        : False

ComponentId           : 77877df3-50e6-4f1e-87d6-60752ceeac8d
TopologyId            : f2400c8e-57e9-4253-9b15-a3c8f635272e
ServerId              : afc3e3f4-0bc5-49a7-8935-e55ab43cad3b
Name                  : CrawlComponent0
ServerName            : demo-sp
ExperimentalComponent : False

ComponentId           : 88573eec-df16-4fb4-a09a-5919ec97a5f1
TopologyId            : f2400c8e-57e9-4253-9b15-a3c8f635272e
ServerId              : afc3e3f4-0bc5-49a7-8935-e55ab43cad3b
Name                  : AdminComponent1
ServerName            : demo-sp
ExperimentalComponent : False

ComponentId           : fa7289e8-26ad-475c-9ab9-2860fbaf29cf
TopologyId            : f2400c8e-57e9-4253-9b15-a3c8f635272e
ServerId              : afc3e3f4-0bc5-49a7-8935-e55ab43cad3b
Name                  : ContentProcessingComponent1
ServerName            : demo-sp
ExperimentalComponent : False

ComponentId           : 67dd1bb3-5f24-473f-a2a1-881da706e83a
TopologyId            : f2400c8e-57e9-4253-9b15-a3c8f635272e
ServerId              : afc3e3f4-0bc5-49a7-8935-e55ab43cad3b
Name                  : QueryProcessingComponent1
ServerName            : demo-sp
ExperimentalComponent : False

ComponentId           : 50b01805-8126-4067-9ed6-07739848f630
TopologyId            : f2400c8e-57e9-4253-9b15-a3c8f635272e
ServerId              : afc3e3f4-0bc5-49a7-8935-e55ab43cad3b
Name                  : CrawlComponent1
ServerName            : demo-sp
ExperimentalComponent : False
```

3. Remove the object by using the following command:

```
Remove-SPEnterpriseSearchServiceComponent -Identity <ComponentID> -
SearchTopology $clone
```

With that, the search component has been removed from the cloned topology.

Activating a search topology

After making modifications to the cloned search topology, you need to activate the cloned topology. To do so, use the following syntax:

```
Set-SPEnterpriseSearchTopology -Identity $clone
```

The updated search topology is now active.

In the next section, we'll review the troubleshooting of performance issues.

Troubleshooting Search

Search issues can be broken down into two areas: performance and unexpected results. In this section, we'll look at some steps and resources for helping deliver good performance from your search environment, as well as ensuring that the delivered search results reflect the expected search results.

Performance

Search performance is a function of the disk, processor, and memory constraints of the servers performing either crawls or retrieving results. From a crawling perspective, enabling *continuous crawling* is a resource-intensive task. If you have enabled this on one or more content sources, you may need to scale the resources or number of servers using one of the predefined architectures under the *Configuring search farms* section of this chapter. You'll need to plan for the number of items you have in your corpus. If you're experiencing slow search and crawl performance with your current deployment, you'll likely need to increase the number of servers running search components.

You can review crawling performance with **Crawl Health Reports**. On the **Search Service** administration page, select **Crawl Health Reports**. The following screenshot shows an example of the CPU and memory load graph for a server that may be overutilized (or needs additional resources to be added to it):

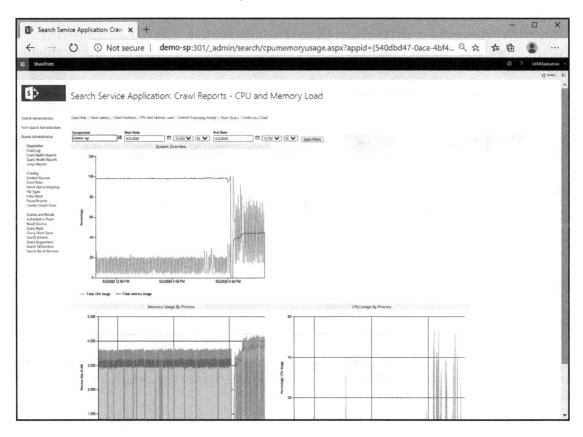

The logs and performance graphs can be used for resource planning and helping to determine where performance issues exist.

Crawl completeness

If search results don't contain information that should be there, you may have an error that needs to be resolved with the crawler or the content that has been placed in a site. To do this, you can search the crawl log.

You can navigate to the crawl log from **Central Administration (Central Administration | Application Management | Manage Service Applications | Search Service Application | Crawl Log)**:

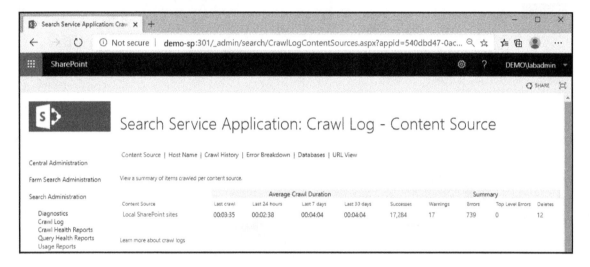

The log will show results for successful, warning, error, and deletion actions. You can click on the number in the related column for a content source to view the details. In the following screenshot, you can see various warnings:

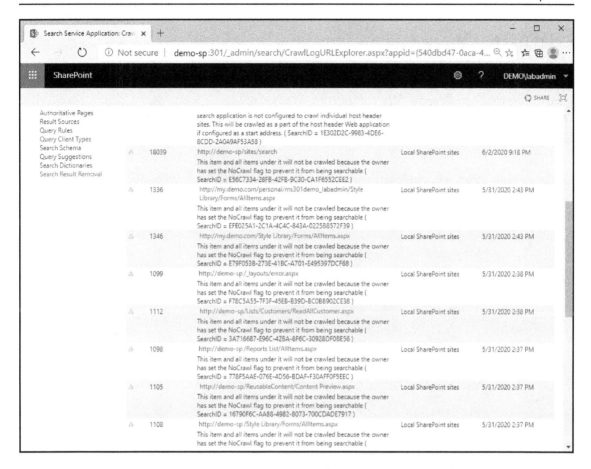

Select the appropriate link to retrieve the individual document or URL returning the error. Warnings and errors may be generated for many types of issues, including performance, data corruption, or security.

Summary

In this chapter, we covered how to configure and manage queries and result sources, as well as the search schema and the different ways they can be used in both SharePoint classic and modern designs. We configured enterprise search features, including taxonomy and keywords, and examined the properties and options for the enterprise search web parts. Finally, we spent some time reviewing the report generation capabilities of the service.

In the next chapter, we will learn about the monitoring and service capabilities of the SharePoint Online service.

Exploring Office Service Applications

9

In this chapter, we're going to dig into some of the SharePoint Office service applications. These service applications are not installed or configured by default. Instead, they are used primarily by developers to build automation services for converting and rendering Office-based documents.

We'll cover the following topics:

- Exploring Office service applications
- Enabling Excel Services
- Enabling PowerPoint Automation Services
- Enabling Visio Graphics Service
- Enabling Word Automation Services

By the end of this chapter, you should have an understanding of the basic Office service applications, how they can be used by developers in your organization, and how to prepare them for use.

Let's go!

Exploring Office service applications

As we mentioned in the introduction, Office service applications are primarily used by developers to extend the capabilities of the Office platform into SharePoint Server. The core Office service applications are as follows:

- **Access Services**: Enables users to create and manage web-based Microsoft Access applications.
- **Excel Services**: Enables users to work with Excel workbooks on SharePoint sites.

- **Machine Translation Service**: Provides language translation services for sites and documents.
- **PerformancePoint Service**: This is used to create and publish dashboards.
- **PowerPoint Automation Services**: This enables unattended conversion of PowerPoint presentations.
- **Visio Services**: Enables users to interact with Visio drawings on SharePoint sites.
- **Word Automation Services**: This enables unattended conversion of Word documents.

Microsoft is de-emphasizing Access and Excel services and recommends that organizations use newer cloud-based technologies like Power Automate, Power Apps, and Azure SQL services for data-driven application development and rendering. Microsoft is also recommending using Power BI for dashboards or visualizations instead of Performance Point, where possible. Finally, Microsoft has deprecated the Machine Translation Service, though it does remain supported.

As Microsoft is recommending replacements for Access, Excel, and Machine translation services, they will not be specifically addressed here.

 Many of the Office service applications require Office Online Server. Deploying and configuring Office Online Server is discussed in `Chapter 12`, *Implementing Hybrid Teamwork Artifacts*.

In the next section, we'll look at enabling Excel Services.

Enabling Excel Services

Many of the features in Excel Services for rendering documents have been included in Office Online Server. To enable the integration of Excel Services integration with Office Online Server, follow these steps:

1. Following the guidance in `Chapter 12`, *Implementing Hybrid Teamwork Artifacts*, ensure that Office Online Server has been installed and configured on a separate server.
2. On a SharePoint Server in the farm, launch an elevated SharePoint Management Shell instance.

3. Run the following command, replacing <oos-server> with your Office Online server:

```
$Farm = Get-SPFarm
$Farm.Properties.Add("WopiLegacySoapSupport","http://<oos-server>/x
/_vti_bin/ExcelServiceInternal.asmx")
$Farm.Update()
```

Adjust any other parameters as necessary using the Set-OfficeWebAppsFarm cmdlet on the Office Online Server computer. Commonly updated parameters include setting the ExcelWarnOnDataRefresh value to $false (to prevent repeated prompts when updating workbooks with external content or PowerPivot data) and increasing the maximum workbook size to 100 MB or greater using the ExcelWorkBookSizeMax parameter.

Next, we'll look at enabling PowerPoint Automation Services.

Enabling PowerPoint Automation Services

In Chapter 3, *Managing and Maintaining a SharePoint Farm*, you learned how to enable service applications. However, the PowerPoint Automation Services application is *not* listed:

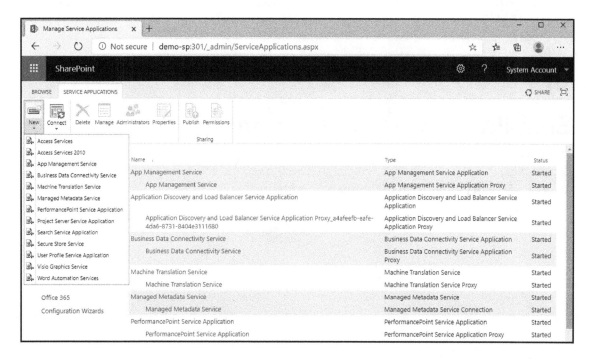

The PowerPoint Automation Services application must be enabled via the SharePoint Management Shell (if it was not already enabled during `MinRole` setup). To configure it, follow these steps:

1. Launch an elevated SharePoint Management Shell.
2. Run the following command to create the service application instance:

   ```
   New-SPPowerPointConversionServiceApplication -Name "PowerPoint
   Automation Service Application" -ApplicationPool "SharePoint Web
   Services Default"
   ```

3. Run the following command to create the service application proxy instance:

   ```
   New-SPPowerPointConversionServiceApplicationProxy -Name "PowerPoint
   Automation Service Application Proxy" -ServiceApplication
   "PowerPoint Automation Service Application" -AddtoDefaultGroup
   ```

4. On the servers hosting the PowerPoint Automation Services application, create a new folder
   at `C:\ProgramData\Microsoft\SharePoint\PowerPointConversion`.

Once this has been completed, PowerPoint Automation Services is ready for use.

You can learn more about interacting with the PowerPoint Automation Services API at `https://docs.microsoft.com/en-us/sharepoint/dev/general-development/powerpoint-automation-services-in-sharepoint`.

In the next section, we'll review Visio services.

Enabling Visio Graphics Service

The Visio Service service application allows users to load, display, and interact with Visio documents (`.vsdx`, `.vsdm`, and `.vdw` files). Users have the ability to publish diagrams with the Visio desktop application to the SharePoint farm.

Additionally, Visio services can refresh the following data types that are configured using the Visio application:

- SQL Server
- Excel workbooks (if the data is hosted on the same farm)
- SharePoint Server lists (if the data is hosted on the same farm)
- External lists configured through Business Connectivity Services
- OLEDB or ODBC
- Custom data providers that have been implemented as .NET Framework assemblies

This data integration allows Visio services to refresh documents automatically when they are opened.

To configure Visio Graphics Service, follow these steps:

1. Launch **Central Administration**.
2. Under **Application Management**, select **Manage service applications**:

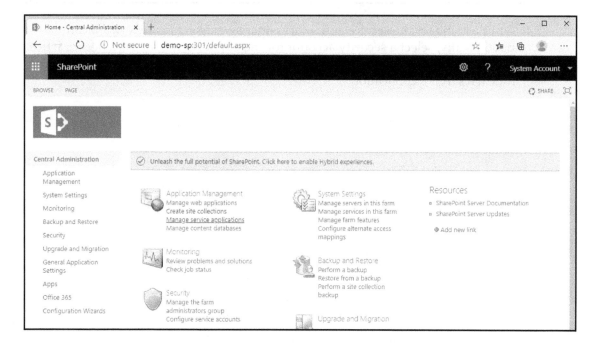

3. Click **New** | **Visio Graphics Service**:

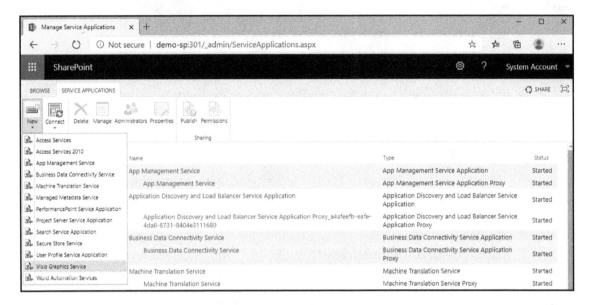

4. Enter a name for the service and select either an existing (or new) application pool. Finally, select a security account for the application pool. As a best practice, Microsoft typically recommends creating a separately managed service account for each service application:

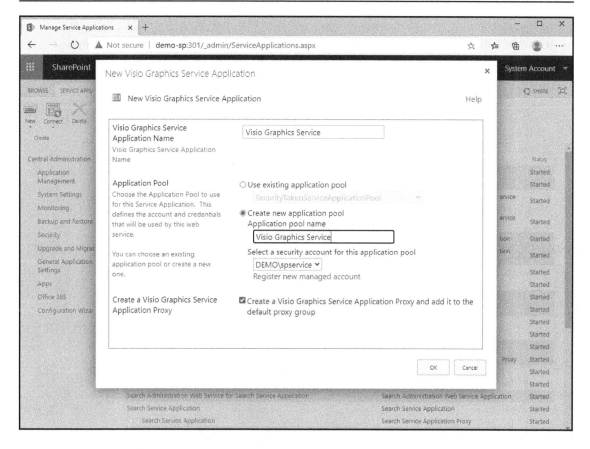

5. Click **OK**.

After the service has been added, you can refresh the **Manage Service Applications** page to display the newly configured Visio Graphics Service.

Users can publish diagrams to SharePoint using the **File | Save** dialog box in the Visio application, entering the URL of a SharePoint site, as shown in the following diagram:

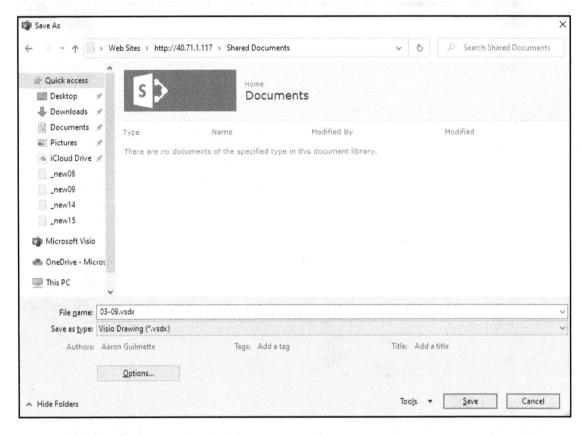

Visio diagrams can then be displayed by configuring a Visio web part on a site page. Visio documents can also be rendered using Visio Web Access by simply opening a drawing directly from a document library:

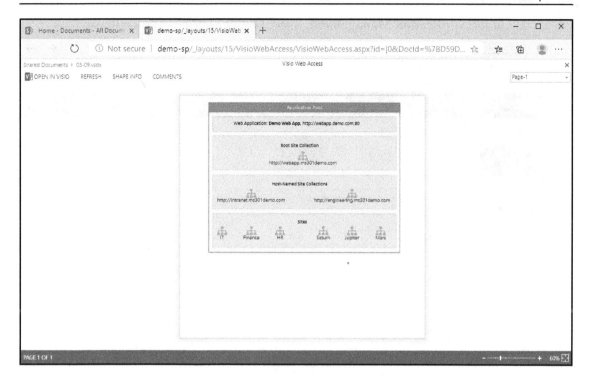

In the next section, we'll configure Word Automation Services.

Enabling Word Automation Services

Like PowerPoint Automation Services, Word Automation Services allows you to programmatically convert document formats supported by Microsoft Word. To enable the Word Automation Services application, follow these steps:

1. Launch **Central Administration**.

2. Under **Application Management**, select **Manage service applications**:

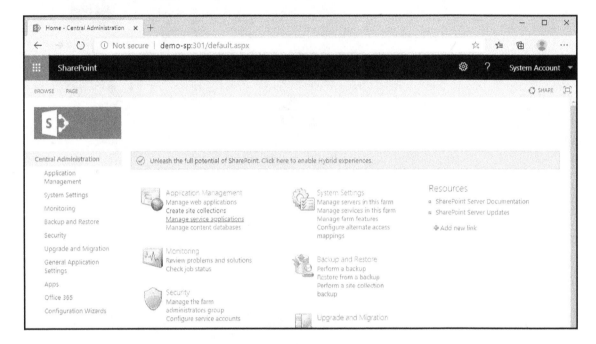

3. Click **New** | **Word Automation Services**:

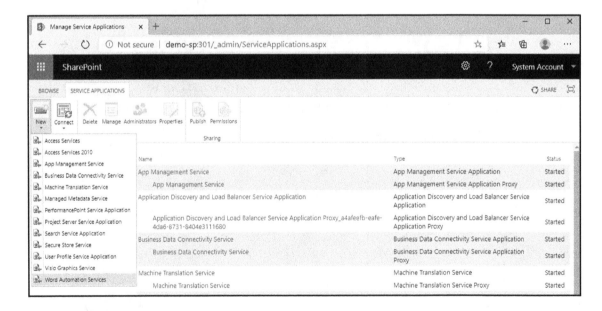

4. Enter a name for the service and select either an existing (or new) application pool. Finally, select a security account for the application pool. As a best practice, Microsoft typically recommends creating a separately managed service account for each service application:

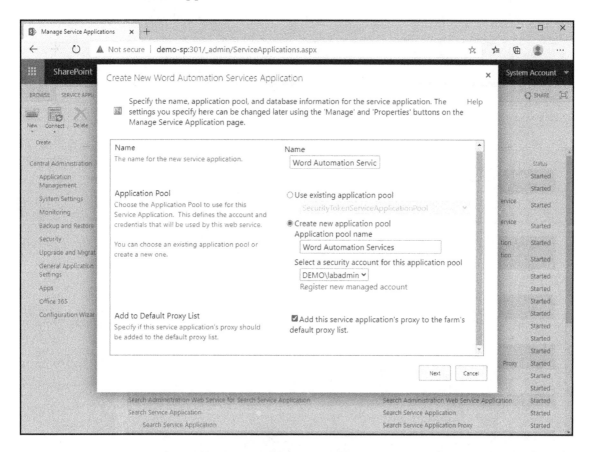

5. Click **Next**.

6. Enter the name of the server that will be hosting the SQL database required for Word Automation Services, and then enter the name that you wish to use for the new database. It's recommended that you use Windows authentication to the database, but you can use SQL authentication if you provide credentials:

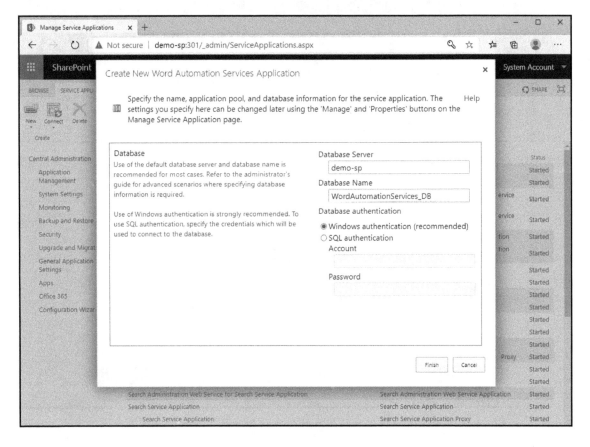

7. Click **Finish**.

The Word Automation Services service application will be configured, and you should be redirected back to the **Manage Service Applications** page. The Word Automation Services service application is now available for use. To configure or update parameters for the Word Automation Services service application, click the **Word Automation Services** link on the **Manage Service Applications** page. You can use the properties page for the service, as shown in the following screenshot:

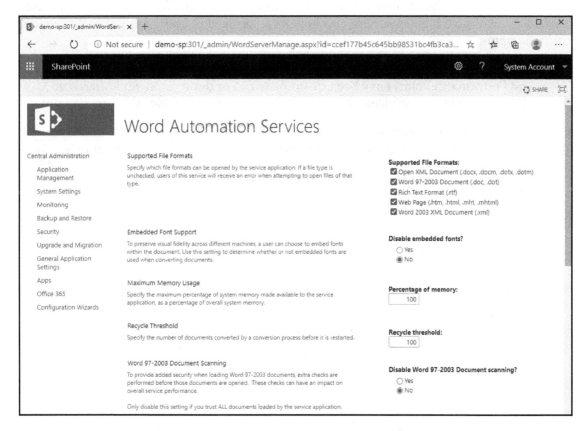

The Word Automation Services service application is now ready to use. For information on developing interfaces to Word Automation Services, see `https://docs.microsoft.com/en-us/sharepoint/dev/general-development/what-s-new-in-word-automation-services-for-developers`.

Summary

In this chapter, you learned about the basic Office service applications that are either available as part of Office Online Server or as native service applications in SharePoint Server 2019, such as PowerPoint and Word Automation Services, as well as Visio Graphics Service. PowerPoint and Word Automation Services can be used by developers to programmatically update or convert supported documents in document libraries. You also learned about Visio Graphics Service and how it can be used to both manipulate and render Visio diagrams.

Finally, it's also important to note that some Office application services (such as Access, Excel, and PerformancePoint) have been de-emphasized or deprecated, and Microsoft recommends using newer services available in Microsoft 365 to replace this functionality.

In the next chapter, we'll begin exploring the concepts around various SharePoint Hybrid scenarios and how Microsoft 365 can be used to further enhance your organization's collaboration capabilities.

Overview of SharePoint Hybrid 10

SharePoint Online is a cloud-based collaboration platform, a corollary to SharePoint Server. Just like SharePoint Server, it comprises sites, storage, and apps. In a SharePoint hybrid configuration, you can connect your on-premises SharePoint deployment to Office 365 to extend your infrastructure into the cloud. As a best practice, Microsoft recommends planning new workloads in Office 365, so configuring your existing environment to integrate with SharePoint Online is important.

SharePoint Hybrid has several core components:

- **Hybrid Sites features**: Redirect SharePoint Server OneDrive or My Sites to Office 365 OneDrive for Business and configure hybrid user profiles and hybrid site following.
- **Hybrid search**: Enable users to search for content both on-premises and on SharePoint Online.
- **Hybrid app launcher**: Configure the Office 365 app launcher to help users navigate between on-premises and online environments.
- **Hybrid taxonomy**: Extend managed metadata and content types from SharePoint Server into SharePoint Online.
- **Business-to-Business (B2B) extranet sites**: Create members-only partner sites on the Office 365 platform to allow external users access to relevant online content.
- **Hybrid self-service site creation**: Configure self-site creation tasks to a corresponding SharePoint Online process.

In this chapter, we're going to discuss, at a high level, how to determine what SharePoint Hybrid solutions are right for your organization by means of the following approaches:

- Evaluating hybrid scenarios
- Planning for hybrid service integration

By the end of this chapter, you should have an understanding of the high-level capabilities of SharePoint hybrid scenarios and a framework for evaluating which services to deploy.

Let's go!

Evaluating hybrid scenarios

Any deployment plan should begin with business goals—you need to understand what the goals are so that you'll be able to see whether you're successful. In evaluating SharePoint Hybrid scenarios, you need to understand what SharePoint Hybrid features are available, what benefits they bring, and whether they are valuable to your organization.

Examples of mapping business goal requirements may include the following:

- Maybe your organization is trying to avoid investing in more on-premises storage for home directories, so transitioning on-premises file storage to OneDrive for Business is a possible solution.
- If you are extending your presence into Office 365 and are concerned about users being able to easily find data, configuring hybrid search can help users locate content on either platform.

The following sections will help you evaluate the capability of the features.

Sites

Sites are the primary content component of SharePoint Server. There are several site-centric features that are possible with SharePoint Hybrid, including configuring redirection, site and document following, and self-service site creation.

The following features are included as part of Hybrid Sites:

- Site following
- OneDrive for Business
- Self-service site creation
- App launcher

Site following

In a SharePoint Server environment, users can follow sites and documents. Hybrid sites enable a consolidated site-following view in SharePoint Online, bringing together the lists of sites followed from both SharePoint Server and SharePoint Online. While this feature doesn't *migrate* the on-premises followed-sites list to Office 365, users can refollow on-premises sites and the followed-sites list in Office 365 will update accordingly.

OneDrive for Business

Hybrid OneDrive for Business enables you to shift data consumption from your SharePoint on-premises deployment (if it exists) to OneDrive for Business in Office 365. Hybrid OneDrive for Business enables your users to combine on-premises SharePoint site usage and OneDrive for Business in SharePoint Online.

While configuring hybrid OneDrive for Business does enable you to redirect users accessing their on-premises My Site, it does not migrate or synchronize data.

Configuring Hybrid OneDrive for Business also enables hybrid user profiles. Hybrid user profiles redirect on-premises profile views to the user's Office 365 profile.

Self-service site creation

Configuring hybrid self-service site creation redirects users from the on-premises site creation tools to the self-service Office 365 group configuration. An Office 365 group (previously known as a modern group) is a construct that combines a group, a group mailbox, a cloud-based OneNote service, and a SharePoint team site as one unit.

App launcher

The extensible hybrid app launcher enables you to create a more seamless navigation experience for users moving between SharePoint on-premises and SharePoint Online environments. The app launcher experience exposes Office 365 apps and custom tiles through the on-premises SharePoint Server app launcher interface.

Search

After Sites, search is the most commonly used SharePoint feature. SharePoint hybrid search has two options available: cloud hybrid search and hybrid federated search.

Cloud hybrid search is the simplest hybrid search option to configure. A search application crawls on-premises content and stores the search index along with cloud content in SharePoint Online. This single index exposes all SharePoint content (both on-premises and what is already in SharePoint Online) to the Microsoft Graph and search services. Results are displayed and ranked based on their relevance, regardless of the source of the content, and presented in a single result block. Additionally, users can filter searches to only include online or on-premises content. Users access cloud hybrid search from SharePoint Online.

For most organizations, cloud hybrid search is recommended. Cloud hybrid search has the following advantages:

- Users can see unified search results from multiple sources.
- Cloud hybrid search doesn't require an upgrade of the SharePoint farm environment to 2016 or 2019.
- No on-premises upgrades are required for the search index.
- There is a lower total cost of ownership for search since no additional on-premises hardware or capacity needs to be deployed moving forward. The enterprise search index is stored in SharePoint Online.
- Microsoft Graph applications, such as Delve, can surface content to users.
- It is simpler to deploy, maintain, and troubleshoot.

The hybrid federated search returns content from two separate indices (Office 365 and on-premises SharePoint Server). Results are grouped and ranked independently according to their source and then displayed in separate result blocks, whereas cloud hybrid search uses a single result block. Hybrid federated search may be a good option when your organization has potentially sensitive content that it wants to keep separate from Office 365, but still wants users to be able to search for it from a single plane of glass.

Microsoft recommends configuring cloud hybrid search, although you might wish to implement hybrid federated search or use a combination of hybrid federated search and cloud hybrid search (such as for sensitive content sets or unavailable features). Microsoft recommends cloud hybrid search for most customers, though it's important to know when to use which.

When planning a hybrid search deployment, it's important to understand what features are different or unavailable, or what technologies and features have superseded what you currently have deployed. With an adoption and change management mindset, it's also important to focus on usability and the best end user experience.

From an on-premises planning perspective, when configuring cloud hybrid search, the on-premises SharePoint server that hosts the cloud search service application needs at least 100 GB of storage space. From a cloud storage planning perspective, SharePoint Online can index 1 million items in every 1 TB of space.

Taxonomy

Hybrid taxonomy allows you to define a single SharePoint taxonomy to span on-premises SharePoint Server and SharePoint Online environments. The benefit is that you can use a single metadata set between both platforms.

Unlike other hybrid configurations, the taxonomy is different in that the shared component is mastered online. With other hybrid solutions (such as hybrid identity with Azure **Active Directory** (**AD**) or Exchange Hybrid), the on-premises system is the source of truth or authority, and the cloud receives a synchronized copy of this on-premises environment.

When you configure hybrid taxonomy and content types, your on-premises term store configuration and content types are copied to SharePoint Online. A timer job is then configured on-premises to update the local taxonomy and content types. Moving forward, taxonomy and content types should only be updated in SharePoint Online.

B2B extranet sites

An extranet, generally, is a resource that is available to people outside an organization for the purposes of sharing data. In SharePoint terminology, an extranet is a restricted site or site collection that enables your organization to share information with external users while protecting internal assets. With a SharePoint Hybrid configuration, you can direct external users to sites in Office 365 using the B2B hybrid feature.

There are many advantages to configuring extranet sites in SharePoint Online:

- Site collections can be configured to allow all users or owners to invite external vendor or partner users.
- Administrators can limit the list of external domains to which the organization allows sharing.
- Activity and audit reports can be used to track site access and usage.

- External guest users can be restricted to only a single site, preventing access to unauthorized internal resources.
- External guest users can be restricted to only being able to accept invitations and sign in to SharePoint Online from the address that received the email, thereby preventing further sharing with unapproved parties.

In planning your extranet model, you need to make three core decisions:

- **Invitation model**: This answers the question of *how do users get access?* You can allow either all users or only the extranet site owners to invite others through the user interface, or prevent all users from inviting anyone and manage the process administratively (or with a third-party application).
- **Licensing**: This helps answer the question of *which features can guests have?* By default, SharePoint guest users have limited capabilities governed by the restrictions set for the group into which they are placed. Authenticated external users can use Office Online to view and edit documents, but further features (such as downloading and installing Office 365 ProPlus, being able to create and manage sites, or being able to share resources) require the assignment of additional licenses.
- **Identity management and governance**: Part of the ongoing management of any resource is the governance of identity and access. You can use the identity and access management tools that are part of Azure AD Premium or develop a custom solution.

Now that you have an understanding of the features available in SharePoint Hybrid, let's look at some of the things to consider when planning the integration.

Planning for hybrid service integration

Hybrid feature deployments can provide a path to the cloud for many services and workloads. While you may have learned *what* options are available, you'll still need to work with business stakeholders to determine whether the features are valuable. These stakeholders will depend on the services being deployed, the types of applications supported, the location of other business data assets, and the impact on users. They may include directors and other executives, service desk managers, call-center representatives, or other users. Also, you may need to engage with people such as enterprise architects and application integration specialists to determine what (if any) portions of your SharePoint Server platform are well suited for a hybrid topology.

If your organization has developed custom solutions or forms, configured integrations with other on-premises applications or data sources, or deployed plugins and workflows to interact with other on-premises applications, you'll need input and direction from those responsible for managing those solutions.

After you have identified the organizational goals and capabilities of the platforms, it's important to become familiar with the process and prerequisites of any implementation. SharePoint hybrid features have several dependencies, as the following list indicates:

- Configuring any hybrid service scenario that includes SharePoint Server 2013 requires SharePoint Server 2013 Service Pack 1 at a minimum. As a general rule, all SharePoint environments should be configured with the most recent updates. For a list of available updates, see `https://docs.microsoft.com/en-us/officeupdates/sharepoint-updates`.
- My Sites must be configured for Hybrid OneDrive.
- The Subscription Settings service application must be configured for Hybrid Sites features.
- The User Profile service must be configured to synchronize from AD, either via SharePoint AD Import or Microsoft Identity Manager.
- At least one Managed Metadata Service application must be configured.
- The App Management service application must be configured.
- You must have already configured Azure AD Connect for identity synchronization between the AD and Microsoft Azure AD. For more information on configuring Azure AD Connect, refer to `https://docs.microsoft.com/en-us/azure/active-directory/hybrid/`.

Finally, if your current SharePoint Server environment is under-utilized, very outdated, or hasn't undergone a lot of customization, it may be more worthwhile to investigate whether migrating content directly makes more business sense. SharePoint Online migration options are discussed in `Chapter 14`, *Overview of the Migration Process*, and `Chapter 17`, *Migrating Data and Content*.

Summary

In this chapter, you learned about the capabilities of various SharePoint Hybrid scenarios, including Hybrid Sites, taxonomy, B2B extranets, and search. Configuring these features will allow an organization to extend into the SharePoint Online platform at their own pace and position features in a way that best helps drive successful adoption by end users. Most importantly, you learned about the importance of considering the interests of business owners and stakeholders as part of a SharePoint Hybrid strategy.

In the next chapter, we'll begin configuring common SharePoint Hybrid features.

Planning a Hybrid Configuration and Topology

11

As the online services collaboration landscape continues to evolve, many organizations will start shifting workloads away from on-premises resources. However, for larger organizations with significant investments in on-premises architecture, applications, and services, these changes must happen gradually.

Hybrid configurations for SharePoint allow organizations to begin leveraging the newest cloud-based technologies without forfeiting on-premises investments.

In Chapter 10, *Overview of SharePoint Hybrid*, we discussed some of the features of the various SharePoint Hybrid configurations. As a review, SharePoint Hybrid supports the following features:

- **The Hybrid site features**: Redirects SharePoint Server OneDrive or My Sites to Office 365 OneDrive for Business and configures hybrid user profiles and hybrid site followings
- **Hybrid search**: Enables users to search for content both on-premises and on SharePoint Online
- **Hybrid app launcher**: Configures the Office 365 app launcher to help users navigate between on-premises and online environments
- **Hybrid taxonomy**: Extends managed metadata and content types from SharePoint Server into SharePoint Online
- **Business-to-Business (B2B) extranet sites**: Creates members-only partner sites on the Office 365 platform to allow external users access to relevant online content
- **Hybrid self-service site creation**: Configures self-site creation tasks to a corresponding SharePoint Online process

In this chapter, we're going to review some general best practices and the prerequisites for hybrid connectivity with a SharePoint Server 2019 farm, and explore the Hybrid Configuration Wizard.

Planning for Hybrid best practices

Deploying SharePoint Hybrid requires planning from both the business and technology perspectives. Any solution deployed should be vetted against the business requirements, such as application availability, security, integration with the existing line-of-business applications, and end user adoption management.

Technology best practices include the following:

- Ensuring all deployed systems are kept up to date
- Configuring the monitoring and alerting options for services, applications, and system-level notifications
- Implementing security with a least-privileged mindset
- Maintaining adequate documentation of service accounts, web apps, service applications, and site collections
- Implementing change control management
- Maintaining farm, site, and application backups
- Regular testing of disaster recovery procedures
- Identity and security account life cycle management

Each individual Hybrid workload has its own separate recommendations that will be used throughout this book for informing configuration choices and recommendations.

Configuring the prerequisites for a SharePoint Hybrid farm

The SharePoint Hybrid features are configured with a runtime tool—the Hybrid Configuration Wizard. One of the benefits of a web-based runtime tool is that it is always up to date with the latest capabilities. There are a number of prerequisites that need to be met prior to executing the Hybrid Configuration Wizard and connecting to SharePoint Online.

The terms **Hybrid Picker** and **Hybrid Configuration Wizard** are used interchangeably in both the Microsoft documentation and this book. Hybrid Picker was the original name of the configuration tool. Though it was later renamed to Hybrid Configuration Wizard, not all references were updated.

Most requirements are required regardless of the Hybrid services being configured. The generally shared prerequisites for all features are as follows.

For SharePoint Server 2013 only, there is the following:

- **SharePoint Server 2013 Service Pack 1 and the January 2016 public update:** `https://docs.microsoft.com/en-us/officeupdates/sharepoint-updates`

For SharePoint Server 2013, 2016, and 2019, there are the following:

- Microsoft 365 or a SharePoint Online tenant
- Administrative accounts:
    - A cloud account with the global administrator, application administrator, or cloud application administrator roles.
    - The on-premises account must be a member of the SharePoint Farm Administrators group.
    - The on-premises account must have the `securityadmin` server role in the farm's SQL server instance.
    - The on-premises account must be a member of the `db_owner` fixed database role on affected SharePoint databases.
    - The on-premises account must be a member of the local Administrators group on the server where tasks will be performed.
- The Hybrid Configuration Wizard, located at `https://go.microsoft.com/fwlink/?linkid=867176`
- Microsoft Online Services Sign-In Assistant for IT Professionals RTW, located at `https://go.microsoft.com/fwlink/?LinkID=286152`, installed on the search and central administration servers

- The Azure Active Directory module for Windows PowerShell, located at `https:/ /docs.microsoft.com/en-us/powershell/azure/active-directory/install- msonlinev1?view=azureadps-1.0`, installed on the search and central administration servers
- The URL of the SharePoint Online root site collection (`https://<tenant>.sharepoint.com`)

Some hybrid configurations (such as search) may have additional requirements, such as service accounts and user rights assignments. As we explore these configurations in later chapters, those specific requirements will be iterated.

Running the SharePoint Hybrid Configuration Wizard

The SharePoint Hybrid Configuration Wizard can be launched from several different locations, including the following:

- The SharePoint Online admin center
- SharePoint Server Central Administration
- Through a direct download link: `https://go.microsoft.com/fwlink/?linkid= 867176`

The Hybrid Configuration Wizard is a web-based runtime application that downloads the current configuration and capabilities each time it's run. Since the Hybrid Configuration Wizard must be run on a SharePoint server, the easiest method to launch it is from Central Administration:

1. Launch **Central Administration** (SharePoint Server 2016 and 2019) and navigate to **Office 365**, then select **Launch the Hybrid Configuration Wizard**. Alternatively, you can open a web browser directly to SharePoint's Hybrid Picker (`https://go.microsoft.com/fwlink/?linkid=867176`). Both methods will download the newest version of the Hybrid Configuration Wizard:

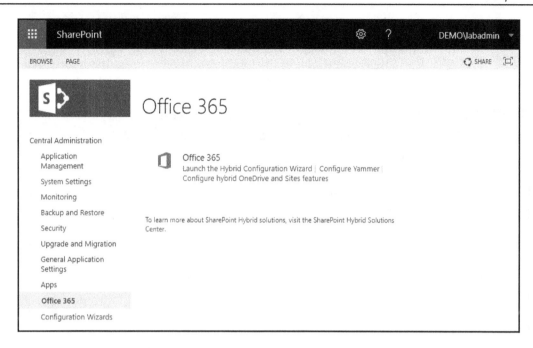

2. Click **Next** to proceed:

3. To connect to the local SharePoint farm, you can use the current user credentials or enter a specific credential. For the Office 365 credentials, enter the username and click **Validate credentials**. In the modern authentication dialog box that follows, enter the account password. After the validation completes, click **Next**:

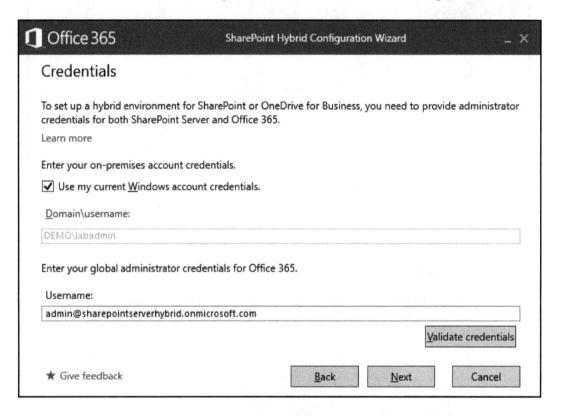

4. Verify that all the prerequisite tests pass. You'll need to resolve any errors before proceeding. Click **Next** when ready:

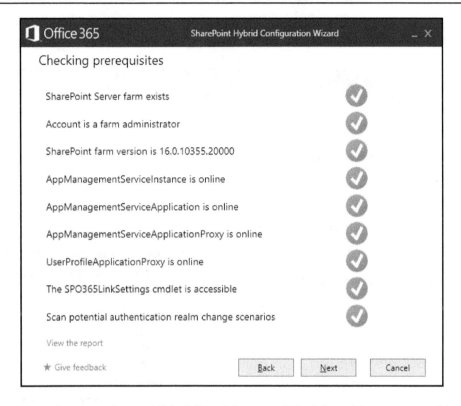

5. If you have configured Workflow Manager (Workflow Manager was discussed in Chapter 3, *Managing and Maintaining a SharePoint Farm*), you'll be prompted to update the security realm to ensure the workflows continue to work. If you don't do this automatically, you'll have to return to the wizard or manually run Set-SPAuthenticationRealm. Click **Next**:

6. Select the options to configure, providing any additional inputs as necessary, then click **Next** to complete the wizard:

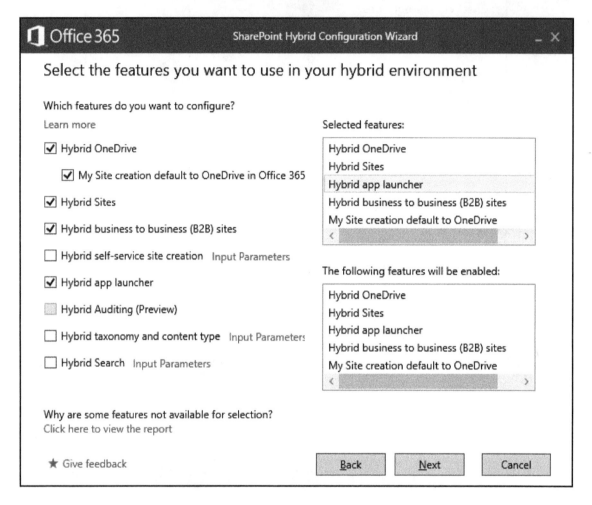

Some of the hybrid features require additional input, such as database or application server names. For options that require additional configuration values, you'll need to click on the feature's corresponding **Input Parameters** link to add the missing values. The setup will not allow you to proceed without the necessary values populated.

In the preceding screenshot, you may have noticed that the **Hybrid Auditing (Preview)** feature is grayed out. This feature was first available on SharePoint Server 2016 deployments but is no longer supported for any version of SharePoint. The capability was deprecated in the November 2019 public update for SharePoint 2016. The **Hybrid Auditing (Preview)** option is grayed out for all versions of SharePoint and will be permanently removed in a future update. For organizations that had originally configured **Hybrid Auditing**, no new data will be sent to SharePoint (regardless of whether or not the November 2019 public update has been applied).

If any additional features are unavailable for your version of SharePoint, you can click on the link to view the report to find out why.

Summary

In this chapter, we reviewed the prerequisites for configuring hybrid connectivity between the SharePoint Server and SharePoint Online environments. We also became familiar with downloading and launching the Hybrid Configuration Wizard. These skills are important as they equip you for beginning to configure hybrid workloads for SharePoint Server.

In the next chapter, we're going to begin configuring hybrid features to enable cross-premises workloads.

Implementing Hybrid Teamwork Artifacts

12

As the online services collaboration landscape continues to evolve, many organizations will be shifting workloads away from on-premises resources. However, for larger organizations with significant investments in on-premises architecture, applications, and services, these changes must happen gradually.

Hybrid configurations for SharePoint allow organizations to begin leveraging the newest cloud-based technologies without forfeiting on-premises investment, allowing organizations to continue utilizing on-premises infrastructure and extending to the cloud where it makes sense. Cloud-based technologies can typically be scaled much more quickly than on-premises resources, so Microsoft recommends looking to the Microsoft 365 platform for new investments.

In this chapter, you'll learn how to configure and manage some of the basic hybrid features. To do this, we'll cover the following topics:

- Hybrid taxonomy and content types
- Hybrid OneDrive for Business
- Hybrid sites
- Hybrid B2B sites
- Hybrid app launcher
- Document rendering for web apps

There are a lot of configuration tasks to get stuck into, so let's go!

Configuring hybrid taxonomy and content types

The hybrid taxonomy feature allows you to extend your on-premises taxonomy into SharePoint Online, enabling a consistent taxonomy in both platforms. Unlike other hybrid features, taxonomy is unique in that once it is synchronized from on-premises to the cloud, it should *only be managed from SharePoint Online*. Hybrid taxonomy was originally released for SharePoint Server 2016 but was later updated so that it was available as far back as SharePoint Server 2013.

Prerequisites

In order to be able to configure hybrid taxonomy and content types, the following prerequisites must be met:

- November 2016 or later public update for hybrid taxonomy (`https://support.microsoft.com/kb/3127940`).
- June 2017 or later public update for hybrid content types (`https://support.microsoft.com/help/3203432`).
- The `Copy-SPTaxonomyGroups`, `SPO365LinkSettings`, and `Copy-SPContentTypes` cmdlets are available.

As a reminder, all hybrid SharePoint features rely on Active Directory account synchronization.

Once the updates have been applied, you can proceed with the configuration.

Updating term store permissions

For the SharePoint Timer job to complete successfully, the Timer service account must be made a member of the **Managed Metadata Service** administrators group. To do this, follow these steps:

1. Launch **SharePoint Management Shell**.
2. Run the following commands:

```
$SPSite = "http://<root site collection">
$SPTimerServiceAccount = (Get-WmiObject win32_service | ? { $_.Name
-eq (Get-Service | ? { $_.Displayname -eq "SharePoint Timer
```

```
Service"}).Name }).StartName
$SPTermStoreName = "Managed Metadata Service"
$Web = Get-SPWeb -Site $SPSite
$TaxonomySession = Get-SPTaxonomySession -Site $Web.Site
$TermStore = $TaxonomySession.TermStores[$SPTermStoreName]
$TermStore.AddTermStoreAdministrator($SPTimerServiceAccount)
$TermStore.CommitAll()
```

Completing this successfully will produce an output on the screen, as shown in the following screenshot:

```
Administrator: SharePoint 2019 Management Shell                                    —   □   ×

PS C:\> $SPSite = "http://demo-sp"
PS C:\> $SPTimerServiceAccount = (Get-WmiObject win32_service | ? { $_.Name -eq (Get-Service | ? { $_.Displayname -eq "S
harePoint Timer Service"}).Name }).StartName
PS C:\> $SPTermStoreName = "Managed Metadata Service"
PS C:\> $Web = Get-SPWeb -Site $SPSite
PS C:\> $TaxonomySession = Get-SPTaxonomySession -Site $Web.Site
PS C:\> $TermStore = $TaxonomySession.TermStores[$SPTermStoreName]
PS C:\> $TermStore.AddTermStoreAdministrator($SPTimerServiceAccount)
PS C:\> $TermStore.CommitAll()
PS C:\> _
```

Next, we'll sync the existing taxonomy with SharePoint Online.

Copying an on-premises taxonomy to SharePoint Online

You need to copy your existing taxonomy to SharePoint Online before running the Hybrid Configuration Wizard. To do so, use the following steps:

1. Launch **SharePoint Management Shell**.
2. Run the following script to copy the non-default taxonomy groups and terms to SharePoint Online:

> The Copy-SPTaxonomyGroups command will fail if your group contains special term sets. In this example, the default groups (People, Search Dictionaries, and System) have been excluded because they contain special term sets that cannot be replicated. If you have additional term sets or the term store is stored in another Managed Metadata Service instance name, you must update those parameters accordingly.

```
$SPOCredential = Get-Credential
$SPOSite = "https://<tenant>.sharepoint.com"
$SPSite = "http://<root site collection>"
```

```
$SPTermStoreName = "Managed Metadata Service"
$Web = Get-SPWeb -Site $SPSite
$TaxonomySession = Get-SPTaxonomySession -Site $Web.Site
$TermStore = $TaxonomySession.TermStores[$SPTermStoreName]
[array]$GroupNames = $TermStore.Groups.Name -notmatch
("People|Search Dictionaries|System")
Copy-SPTaxonomyGroups -LocalTermStoreName $SPTermStoreName -
LocalSiteURL $SPSite -RemoteSiteURL $SPOSite -GroupNames
$GroupNames -Credential $SPOCredential
```

3. Gather a list of the content types you wish to copy to SharePoint Online. To list all of the content types for a particular site, run the following script from **SharePoint Management Shell**:

```
$SPSite = "http://<root site collection>"
$Web = Get-SPWeb -Site $SPSite
[System.Collections.Arraylist]$ContentTypeNames =
$Web.ContentTypes.Name
```

4. Review the values stored in the $ContentTypeNames variable. When you have determined the content types to copy from Office 365, use the following script to copy them. If there are none to exclude, then you can just use the entire $ContentTypeNames collection (however, you'll likely have to remove a lot of built-in things, such as **Health Analyzer Rule Definition**, **Common Indicator Columns**, or other content types that aren't valid in SharePoint Online). You can use the Remove() method to remove them from the $ContentTypeNames variable. Content types will be saved to /sites/contentTypeHub in your SharePoint tenant. If it doesn't exist yet, it will be created:

```
$SPOCredential = Get-Credential
$SPOSite = "https://<tenant>.sharepoint.com"
$SPSite = "http://root site collection>"
Copy-SPContentTypes -LocalSiteUrl $SPSite -LocalTermStoreName
$SPTermStoreName -RemoteSiteUrl $SPOSite -ContentTypeName
$ContentTypeNames -Credential $SPOCredential
```

Now that we know how to clone the taxonomy, let's learn how to run the Hybrid Configuration Wizard.

Running the Hybrid Configuration Wizard

The final step will be to run the Hybrid Configuration Wizard. To do so, follow these steps:

1. Launch **SharePoint Hybrid Configuration Wizard** from the icon on the desktop or from Central Administration.
2. Select the **Hybrid taxonomy and content type** checkbox and then click the corresponding **Input Parameters** link:

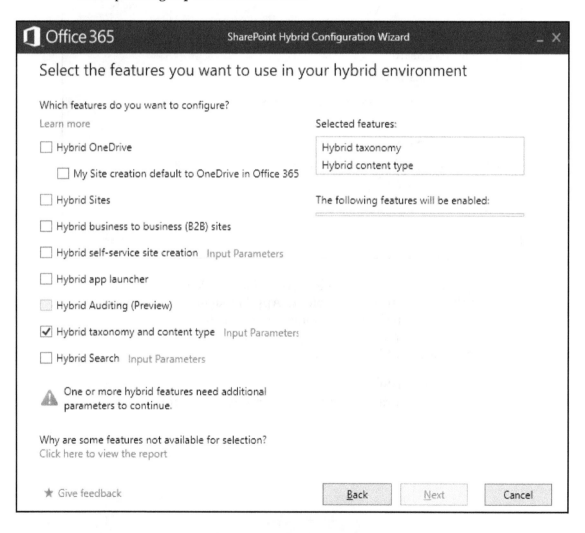

3. On the **Input Parameters** page, enter the local SharePoint site URL, the name of the Managed Metadata Service (the **Local Term Store Name** parameter), and then the names of the groups and content types you want to sync. Then, click **Validate**. If this is successful, click **OK**:

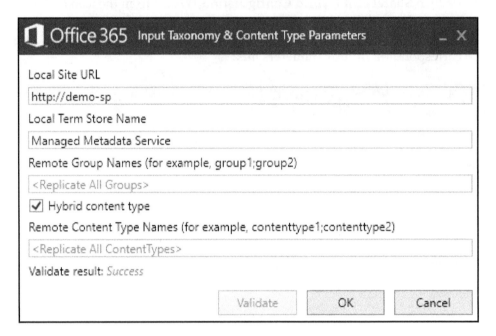

If you don't know the name of your Managed Metadata Service application, you can look refer to **App Management** | **Manage service applications** in Central Administration or run `Get-SPServiceApplication | ? {$_.TypeName -like "*metadata*"}` from SharePoint Management Shell.

4. Click **Next** when you're ready.
5. Review the configuration summary and resolve any errors:

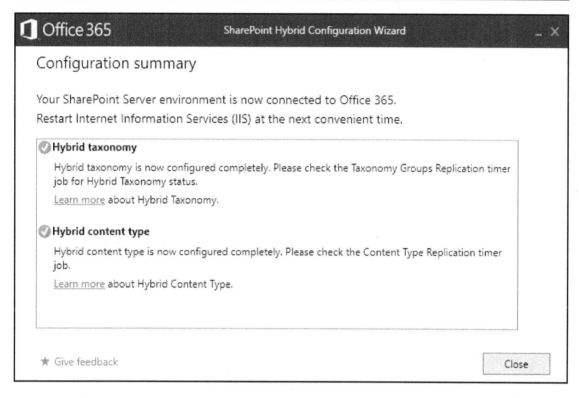

The configuration is complete. Taxonomy and content types should now be managed from SharePoint Online.

Next, we'll look at configuring hybrid OneDrive for Business.

Configuring hybrid OneDrive for Business

Hybrid OneDrive for Business performs redirection for users' **My Sites** area to a corresponding Microsoft 365 OneDrive for Business site. It's important to note that hybrid OneDrive for Business *does not migrate data*.

Configuring hybrid OneDrive for Business can be broken down into four main sections:

- Prerequisites
- Configuring permissions
- Creating a pilot group
- Running the Hybrid Configuration Wizard

Let's take a look at each one.

Prerequisites

To configure hybrid OneDrive for Business and Sites, you must meet the following prerequisites:

- A SharePoint Online license in Office 365
- An administrator account with SharePoint Online admin role privileges
- A SharePoint Online My Sites URL
- An administrator account with membership in the Farm Administrators group

To make hybrid OneDrive services available for your on-premises users, you must subscribe to a Microsoft 365 plan that contains SharePoint Online and then synchronize your on-premises directory to Office 365.

Creating a pilot group

If you decide you want to conduct a pilot of hybrid OneDrive for Business for a small group of users, you can create an audience for your pilot users. Copy and paste the following script into an elevated SharePoint PowerShell console, editing the values for variables such as $MySiteHostUrl, $AudienceName, and $AudienceDescription.

In this example, the script creates an audience where the members are in the IT department.

You can also perform this action in SharePoint Server by going to **Central Administration** and selecting **App Management | Manage Service Applications | User Profile Service | Manage Audiences**, or in SharePoint Online by going to **SharePoint Admin Center | User Profiles | Manage**:

```
$MySiteHostUrl = "https://<sharepoint my sites url>"
$AudienceName = "OneDrive Pilot Users"
$AudienceDescription = "OneDrive Pilot Users"
$AudienceRules = @()
$AudienceRules += New-Object
Microsoft.Office.Server.Audience.AudienceRuleComponent("Department",
"Contains", "IT")
$Site = Get-SPSite $MySiteHostUrl
$ctx = [Microsoft.Office.Server.ServerContext]::GetContext($site)
$AudMan = New-Object Microsoft.Office.Server.Audience.AudienceManager($ctx)
$Audience = $AudMan.Audiences.Create($AudienceName, $AudienceDescription)
$Audience.AudienceRules = New-Object System.Collections.ArrayList
```

```
$AudienceRules | ForEach-Object { $Audience.AudienceRules.Add($_) }
$Audience.Commit()
$Upa = Get-SPServiceApplication | Where-Object {$_.TypeName -eq "User
Profile Service Application"}
$AudienceJob =
[Microsoft.Office.Server.Audience.AudienceJob]::RunAudienceJob(($Upa.Id.Gui
d.ToString(), "1", "1", $Audience.AudienceName))
```

Now that we know how to create a pilot group, let's learn how to configure permissions.

Configuring permissions

To use OneDrive for Business in Microsoft 365, users must have both the **Create Personal Site** and **Follow People and Edit Profile** permissions in Microsoft 365. These are assigned by default, but it's recommended you check you have them.

To verify these permissions, follow these steps:

1. Log in to the Microsoft 365 admin center (`https://admin.microsoft.com`).
2. Expand **Admin centers** and then click **SharePoint**.
3. In the navigation pane, click **More features**, and then click **Open** under **User profiles**:

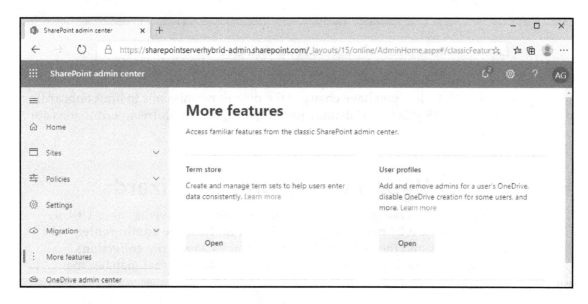

4. Under **People**, click **Manage User Permissions**.

5. On the **Permissions for userprofile_<id>** page, select **Everyone except external users**:

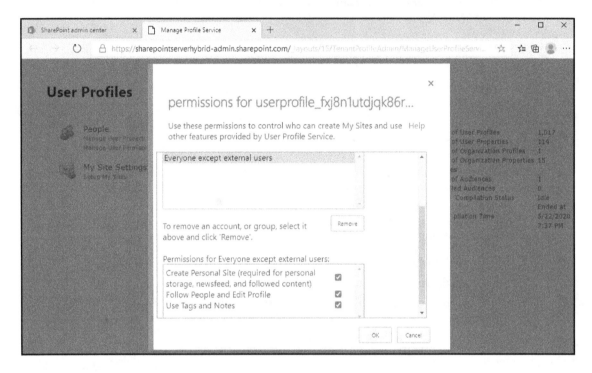

6. Ensure all three boxes are selected and click **OK**.

The **Everyone except external users** group is configured by default, with all the permissions selected. Unless you have changed the default permissions to limit onboarding directly in Microsoft 365, this step is usually just verifying that the correct permissions are already in place.

Running the Hybrid Configuration Wizard

To perform the redirection configuration, you will need the OneDrive/My Sites URL in Office 365. You can locate it by navigating to the **SharePoint Online admin center**, selecting **More features**, and then clicking **Open** under the **Classic site collections page** section. Once it's loaded, you'll need to find the site collection that matches the format `https://<tenant>-my.sharepoint.com`.

When you have your tenant's My Sites URL, you can follow these steps to configure OneDrive for Business redirection:

1. Launch **SharePoint Hybrid Configuration Wizard** from the icon on the desktop or from Central Administration.
2. Only select the **Hybrid OneDrive** checkbox. If you want new My Sites to default to OneDrive for Business, select the sub-option as well:

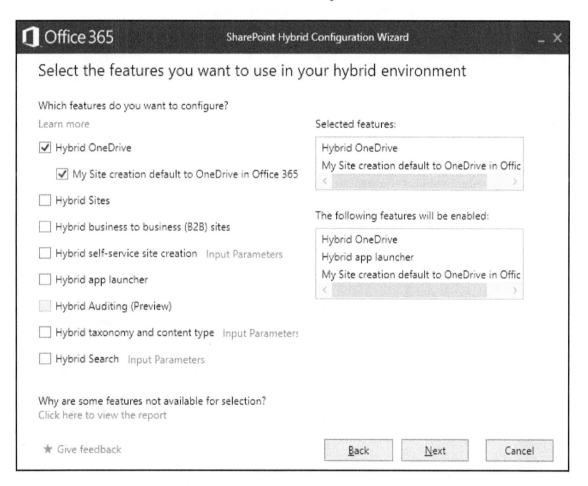

3. Click **Next**.

4. Verify that it has been completed successfully:

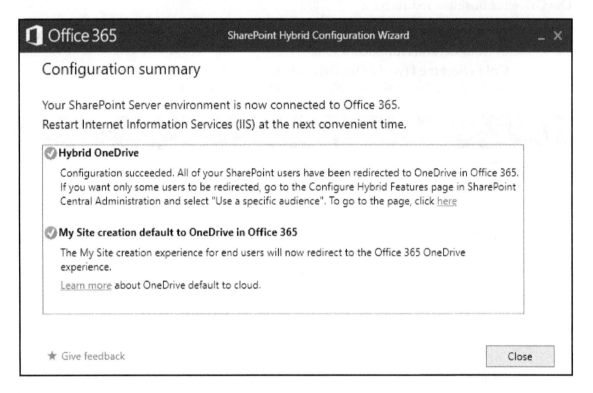

5. If you want to configure **Hybrid OneDrive** for only a specific audience, launch **Central Administration** and select **Office 365**. Then, click **Configure hybrid OneDrive and Sites features**.

6. Select **Use a specific audience**, and then select an audience. Then, click **OK**:

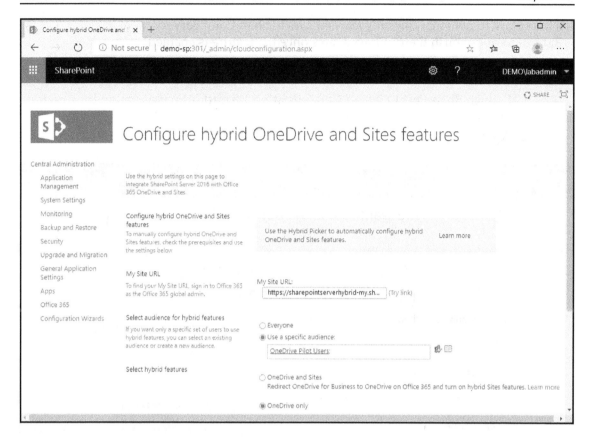

With that, hybrid OneDrive for Business has been configured. Next, we'll look at configuring hybrid sites and their features.

Configuring hybrid sites

With SharePoint, when a user follows a site for updates, it's added to the user's **Followed Sites** list. However, with mixed environments (such as SharePoint Server and SharePoint Online), if users follow sites in both environments, they'll end up with two **Followed Sites** lists containing different items.

With hybrid sites features, the user's sites link from on-premises SharePoint is redirected to SharePoint Online so that users can maintain a single list.

It's important to note that *currently followed sites* in SharePoint Server *are not migrated*. Users will have to re-establish their followed sites once the feature is enabled.

To configure hybrid sites, follow these steps:

1. Launch **SharePoint Hybrid Configuration Wizard** from the icon on the desktop or from Central Administration.
2. After authenticating, select the **Hybrid Sites** option and click **Next**:

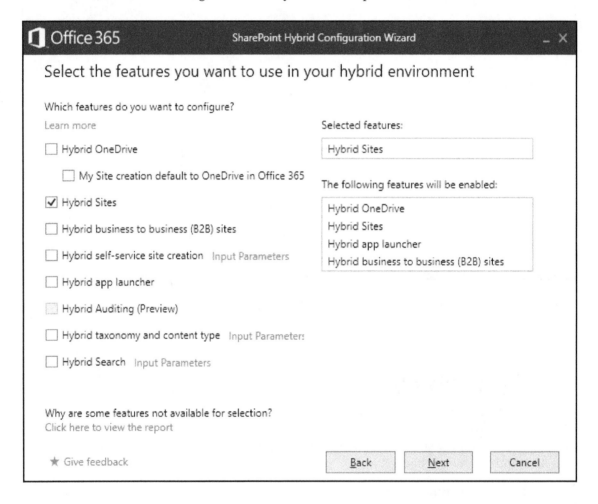

3. Review the configuration summary and click **Close**:

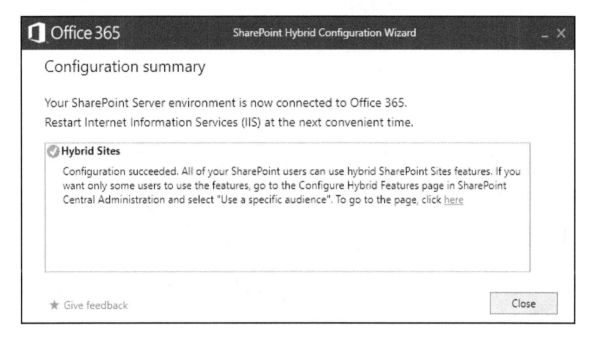

With that, your hybrid sites have been successfully configured. Users can now use the **Follow** link on a site either on-premises or in SharePoint Online and will have a consolidated view of followed sites moving forward.

Next, we'll look at hybrid B2B site configurations.

Configuring hybrid B2B sites

Creating extranet or **business-to-business** (**B2B**) sites in SharePoint removes the need to create and manage Active Directory accounts in an on-premises forest for external users, and also allows organizations to utilize the native site collection security boundaries and Azure security controls present in Office 365.

Some of the benefits of using B2B sites in the Microsoft 365 platform include the following:

- **Low implementation cost**: No additional hardware or software resources outside of your existing Microsoft 365 licensing are necessary; no firewall or other network configurations need to be made.
- **Secure sharing and identity management**: B2B extranet sharing in Microsoft 365 allows us to restrict partners or guests to a single site easily, as well as making it easy for us to manage who can share what and what external domains are allowed or blocked.

- **Collaboration**: Seamless collaboration capabilities allow external guests to interact with internal users via SharePoint and Teams.
- **Data loss prevention**: Here, you can easily apply robust controls to content sharing to prevent privileged data from leaving the organization.
- **Auditing**: Since every activity is logged in Microsoft 365, you have easy-to-access tools that give you visibility into when documents or resources were shared, accessed, and modified.

All of these features make Microsoft 365 and SharePoint Online a compelling choice for B2B sharing.

To configure the Hybrid B2B features using the Hybrid Configuration Wizard, follow these steps:

1. Launch **SharePoint Hybrid Configuration Wizard** from the icon on the desktop or from Central Administration.
2. After authenticating, select the **Hybrid business to business (B2B) sites** option and click **Next**:

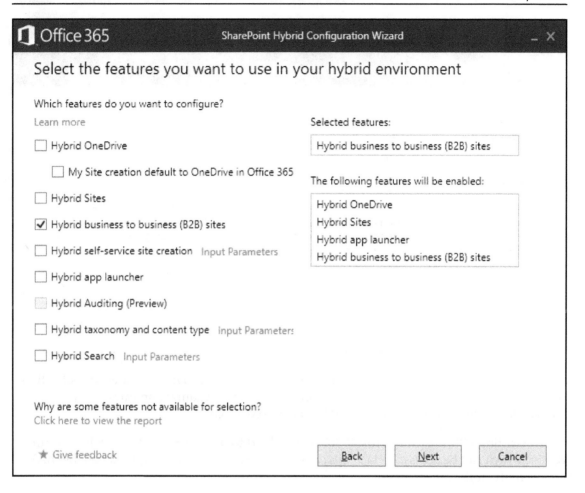

3. Review the configuration summary and click **Close**:

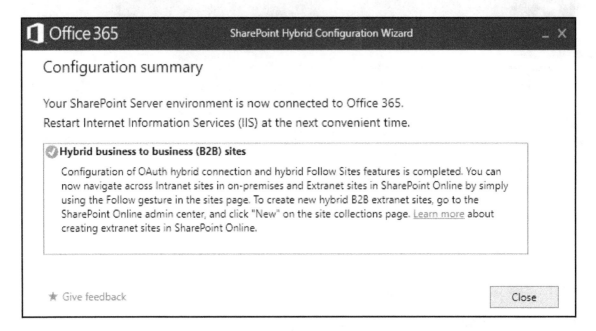

From here, you can now navigate to SharePoint Online and provision a new site collection for use with external guests or connect to another partner organization through Azure Active Directory to create a managed B2B guest experience.

When creating B2B sites in Microsoft 365, you'll need to use the SharePoint Online admin center, which can be accessed via `https://admin.microsoft.com`. To quickly enable B2B connectivity, follow these steps:

1. Launch the Microsoft 365 admin center (`https://admin.microsoft.com`) and navigate to **Admin centers** | **SharePoint**.

2. Navigate to **Sites | Active Sites** and then click **+ Create** to create a site. Fill out the main details and click **Next**:

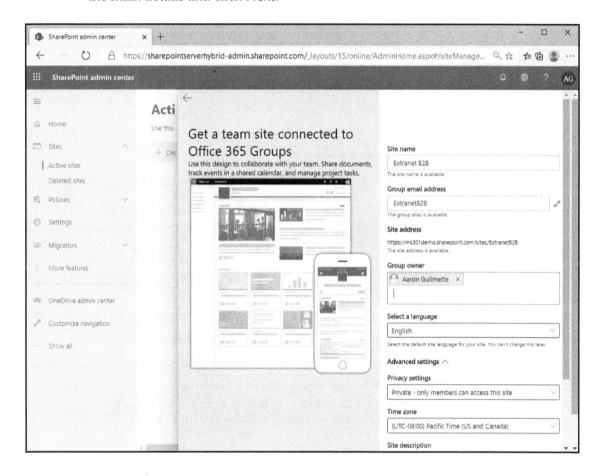

3. Add the email addresses of any internal users or external guests. Click **Finish** when you're done:

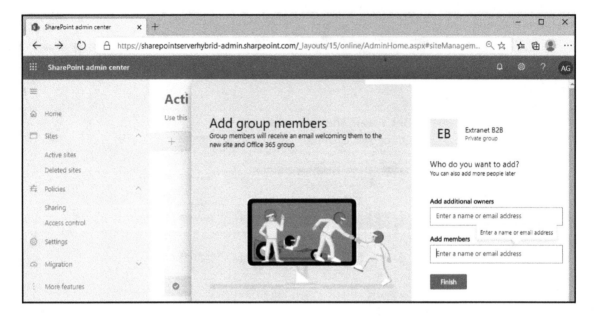

External users will receive an email invitation with a link that they can click on to gain access to the site.

 Creating managed B2B guest experiences is beyond the scope of this book, but you can learn more about this feature at `https://docs.microsoft.com/en-us/microsoft-365/solutions/b2b-extranet?view=o365-worldwide`.

Next, we'll look at the cross-premises integration features of the hybrid app launcher.

Configuring the hybrid app launcher

The hybrid app launcher experience brings down Office 365 integrations into the local SharePoint app launcher. This gives users a single interface for both the SharePoint Server and Office 365 environments.

To configure the hybrid app launcher, follow these steps:

1. Launch **SharePoint Hybrid Configuration Wizard** from the icon on the desktop or from Central Administration.
2. After authenticating, select the **Hybrid app launcher** option and click **Next**:

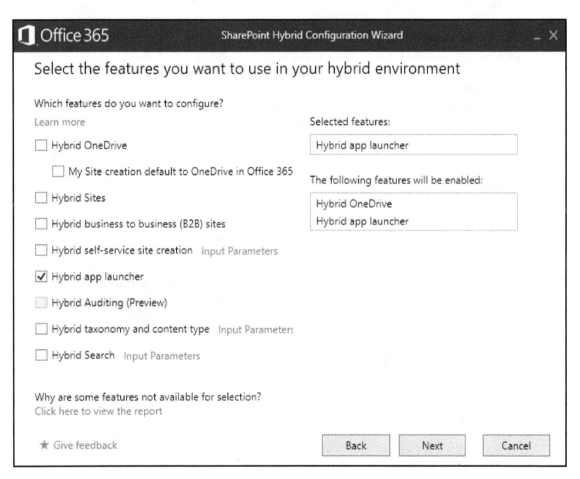

3. Verify that the setup has completed successfully and click **Close**:

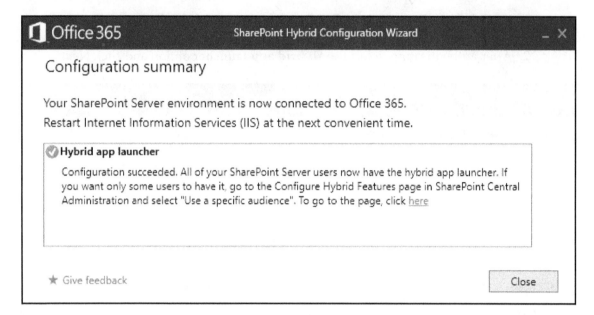

After configuring the hybrid app launcher, you should be able to browse to the local SharePoint farm, click the **App Launcher** icon in the upper left-hand part of the page, and then see links for Office 365 apps integrated with SharePoint on-premises, as shown in the following screenshot:

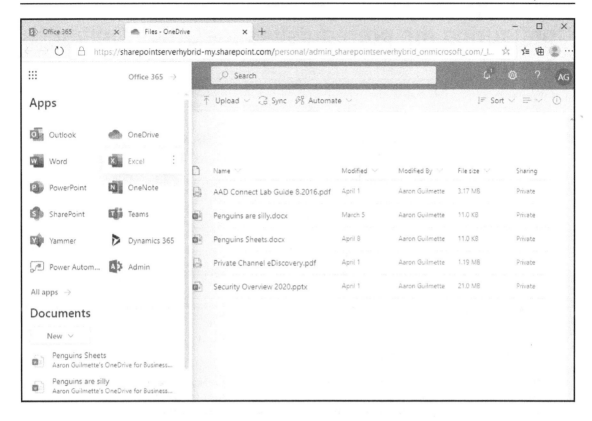

As shown in the preceding screenshot, hovering over the **Excel** icon in the app launcher links us to the Excel online application.

Configuring document rendering for web apps

In Office 365, users have the ability to create and edit documents with the Office Online apps. To enable a corresponding feature on-premises, you need to deploy an Office Online (also known as Office Web Apps) server. Office Online Server is only available through the Volume Licensing Service Center.

In the next few sections, we'll review the prerequisites for installing Office Online Server, as well as configuring it.

Configuring the prerequisites for Office Online Server

Office Online Server has a number of requirements that must be met prior to installation:

- Windows Server 2012 R2 or later.
- Internet Information Services 7.0 or higher with ASP.NET v4.0.
- .NET Framework 4.5.2: `https://go.microsoft.com/fwlink/p/?LinkId=510096`.
- Visual C++ Redistributable Packages for Visual Studio 2013: `https://www.microsoft.com/download/details.aspx?id=40784`.
- Visual C++ Redistributable for Visual Studio 2015: `https://go.microsoft.com/fwlink/p/?LinkId=620071`.
- Microsoft Identity Extensions (`Microsoft.IdentityModel.Extension.dll`): `https://go.microsoft.com/fwlink/p/?LinkId=620072`.
- The target server cannot have any existing SharePoint components installed on it.

If you are installing on Windows Server 2012 R2, run the following PowerShell command to install the prerequisite components:

```
Add-WindowsFeature Web-Server,Web-Mgmt-Tools,Web-Mgmt-Console,Web-
WebServer,Web-Common-Http,Web-Default-Doc,Web-Static-Content,Web-
Performance,Web-Stat-Compression,Web-Dyn-Compression,Web-Security,Web-
Filtering,Web-Windows-Auth,Web-App-Dev,Web-Net-Ext45,Web-Asp-Net45,Web-
ISAPI-Ext,Web-ISAPI-Filter,Web-Includes,InkandHandwritingServices,NET-
Framework-Features,NET-Framework-Core,NET-HTTP-Activation,NET-Non-HTTP-
Activ,NET-WCF-HTTP-Activation45,Windows-Identity-Foundation,Server-Media-
Foundation
```

If you are installing on Windows Server 2016 or later, run the following PowerShell command to install the prerequisite components:

```
Add-WindowsFeature Web-Server,Web-Mgmt-Tools,Web-Mgmt-Console,Web-
WebServer,Web-Common-Http,Web-Default-Doc,Web-Static-Content,Web-
Performance,Web-Stat-Compression,Web-Dyn-Compression,Web-Security,Web-
Filtering,Web-Windows-Auth,Web-App-Dev,Web-Net-Ext45,Web-Asp-Net45,Web-
ISAPI-Ext,Web-ISAPI-Filter,Web-Includes,NET-Framework-Features,NET-
Framework-45-Features,NET-Framework-Core,NET-Framework-45-Core,NET-HTTP-
Activation,NET-Non-HTTP-Activ,NET-WCF-HTTP-Activation45,Windows-Identity-
Foundation,Server-Media-Foundation
```

We'll learn how to install Office Online Server in the next section.

Installing Office Online Server

Once the prerequisites have been met, you can start installing Office Online Server.

To install and configure Office Online Server for document rendering, follow these steps:

1. Prior to installation, download the **Office Online Server Language Packs** files (`https://go.microsoft.com/fwlink/p/?LinkId=798136`) and save it to a local folder.
2. From the Office Online Server media, run `Setup.exe`.
3. Click the **I accept the terms of this agreement** checkbox and select **Continue**.
4. Choose the destination file location and click **Install Now**:

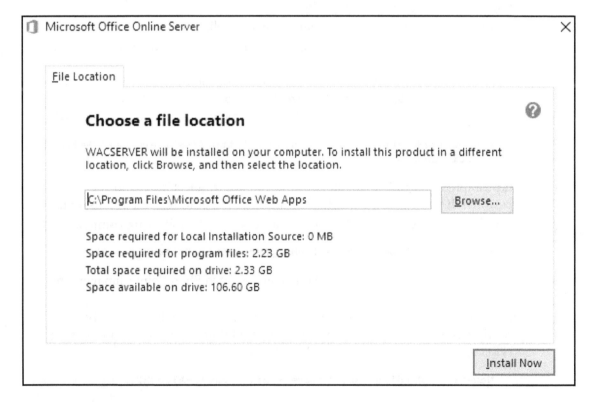

5. When finished, click **Close**.
6. Launch the file you downloaded earlier for the language packs.

With that, the Office Online Server components have been installed. Next, we'll configure the server so that it can integrate with SharePoint Server.

Configuring Office Online Server

After the Office Online Server components have been installed, you will need to configure SharePoint so that it can use them to render or display Office document files. The integration between SharePoint and Office Online Server can use either HTTP or HTTPS communications. HTTP is intended for testing, while HTTPS is intended for production use. HTTPS configuration will require SSL certificates to be deployed.

In this example, we're going to configure a single-server Office Online farm using HTTP.

To do this, follow these steps:

1. In Office Online Server, launch an elevated PowerShell session.
2. Log in to a SharePoint server in the farm.
3. Launch an elevated SharePoint Management Shell.
4. Run the following command, where `<servername>` is the name of the server running Office Online Server. Note that if the `New-OfficeWebAppsFarm` command is not available, you'll need to load the module manually with a command similar to `Import-Module 'C:\Program Files\Microsoft Office Web App\AdminModule\OfficeWebApps\officewebapps.psd1'`:

   ```
   New-OfficeWebAppsFarm -InternalURL "http://<servername>" -AllowHttp
   -EditingEnabled
   ```

5. If prompted to confirm whether editing can occur, select **Y** (as your licensing permits). If your organization has not licensed the editing components, enter **N**.
6. Configure Secure Store to allow HTTP access with the following command:

   ```
   Set-OfficeWebAppsFarm -AllowHttpSecureStoreConnections:$true
   ```

7. Log in to SharePoint Server and launch an elevated SharePoint Management Shell.
8. Run the following command, replacing `<ServerName>` with the newly configured Office Online Server fully qualified domain name. If you're using HTTPS, you can leave off the `-AllowHTTP` parameter:

   ```
   New-SPWOPIBinding -ServerName <ServerName> -AllowHTTP
   ```

9. Update the Office Web Apps zone with the following command:

```
Set-SPWOPIZone -Zone "internal-http"
```

10. Update the SharePoint setting to allow OAuth over HTTP with the following command:

```
$config = (Get-SPSecurityTokenServiceConfig)
$config.AllowOAuthOverHttp = $true
$config.Update()
```

11. Enable the SOAP API for Excel services by adding the WopiLegacySoapSupport property to the SharePoint farm properties, replacing <http://officeonlineserver> with the Office Online Server address:

```
$Farm = Get-SPFarm;
$Farm.Properties.Add("WopiLegacySoapSupport",
<http://officeonlineserver>/x/_vti_bin/ExcelServiceInternal.asmx");
$Farm.Update()
```

With that, Office Online Server has been connected to your SharePoint farm and configured to display and render Office documents.

Next, let's look at some common troubleshooting scenarios.

Troubleshooting hybrid configuration issues

Depending on the complexity of your on-premises SharePoint environment, some issues can arise during configuration. You will need to review these troubleshooting steps to help resolve issues.

Hybrid taxonomy language not found

When configuring hybrid taxonomy and content types, the SharePoint Online tenant must be configured with the same languages as the on-premises environment. If it is not, you'll need to resolve them before moving forward.

In this case, the local languages supported include English and French, while the target SharePoint Online tenant is not configured with French:

```
Administrator: SharePoint 2019 Management Shell                                    —    □    ×
cmdlet Get-Credential at command pipeline position 1
Supply values for the following parameters:
Credential
PS C:\> $SPOSite = "https://sharepointserverhybrid.sharepoint.com"
PS C:\> $SPSite = "http://demo-sp"
PS C:\> $SPTermStoreName = "Managed Metadata Service"
PS C:\> $Web = Get-SPWeb -Site $SPSite
PS C:\> $TaxonomySession = Get-SPTaxonomySession -Site $Web.Site
PS C:\> $TermStore = $TaxonomySession.TermStores[$SPTermStoreName]
PS C:\> [array]$GroupNames = $TermStore.Groups.Name -notmatch ("People|Search Dictionaries|System")
PS C:\> Copy-SPTaxonomyGroups -LocalTermStoreName $SPTermStoreName -LocalSiteURL $SPSite -RemoteSiteURL $SPOSite -GroupN
ames $GroupNames -Credential $SPOCredential
Copy-SPTaxonomyGroups : Language exists in local Term Store but not in remote Term Store: 1036, for more information
please see the online documentation (https://go.microsoft.com/fwlink/?LinkId=798624).
At line:1 char:1
+ Copy-SPTaxonomyGroups -LocalTermStoreName $SPTermStoreName -LocalSite ...
+ ~~~~~~~~~~~~~~~~~~~~~~~~~~~~~~~~~~~~~~~~~~~~~~~~~~~~~~~~~~~~~~~~~~~~~~~~
    + CategoryInfo          : InvalidData: (Microsoft.Share...PTaxonomyGroups:SPCmdletMigrateSPTaxonomyGroups) [Copy-S
   PTaxonomyGroups], SPException
    + FullyQualifiedErrorId : Microsoft.SharePoint.Taxonomy.Hybrid.Cmdlet.SPCmdletMigrateSPTaxonomyGroups

PS C:\> _
```

To resolve this issue, add support for the missing languages in the SharePoint Online Term Store.

Invalid content type name

When copying content types to SharePoint Online, you may run into an error regarding an invalid content type name:

```
Administrator: SharePoint 2019 Management Shell                                    —    □    ×
PS C:\> Copy-SPContentTypes -LocalSiteUrl $SPSite -LocalTermStoreName $SPTermStoreName -RemoteSiteUrl $SPOSite -ContentT
ypeName $ContentTypeNames -Credential $SPOCredential
Begin checking the selected content types.
Found the invalid content type name: Administrative Task
Copy-SPContentTypes : Operation is not valid due to the current state of the object.
At line:1 char:1
+ Copy-SPContentTypes -LocalSiteUrl $SPSite -LocalTermStoreName $SPTerm ...
+ ~~~~~~~~~~~~~~~~~~~~~~~~~~~~~~~~~~~~~~~~~~~~~~~~~~~~~~~~~~~~~~~~~~~~~~~~
    + CategoryInfo          : InvalidData: (Microsoft.Share...eSPContentTypes:SPCmdletMigrateSPContentTypes) [Copy-SPC
   ontentTypes], InvalidOperationException
    + FullyQualifiedErrorId : Microsoft.SharePoint.Taxonomy.Hybrid.Cmdlet.SPCmdletMigrateSPContentTypes
PS C:\> $ContentTypeNames.Remove("Administrative Task")
```

To resolve this conflict, use the `Remove()` method on the `$ContentTypeNames` variable and rerun `Copy-SPContentTypes`.

Site following

If you've configured hybrid site following and a user who does not have an on-premises My Site attempts to follow a new site, you may see an error similar to the following:

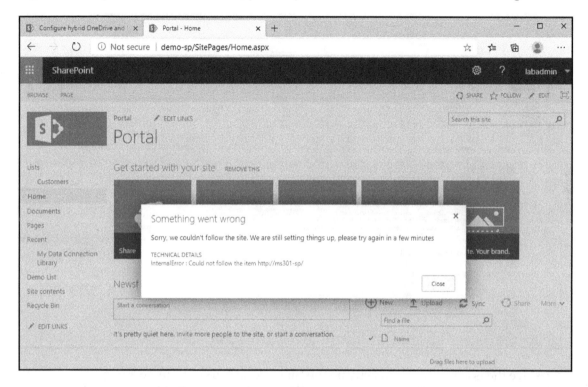

To resolve this issue, provision an on-premises My Site for the user.

Document rendering

If you attempt to configure a new Office Web Apps farm, you may run into the following error:

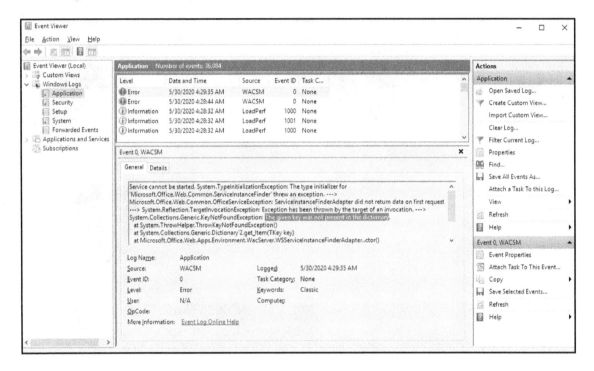

To resolve the issue, ensure you activate the **SharePoint Server Publishing Infrastructure** site collection feature and the **SharePoint Server Publishing** site feature before running `New-OfficeWebAppsFarm`.

Summary

In this chapter, we reviewed the steps we need to take to implement a wide variety of SharePoint hybrid configurations, including sites, extranet or B2B sites, metadata services, the app launcher, and search.

Each of these features brings additional capabilities to help organizations extend their environments into the cloud while maintaining an on-premises footprint.

In the next chapter, we will learn how to implement a hybrid search service application.

13
Implementing a Hybrid Search Service Application

Search is one of SharePoint's core features, as we've seen throughout this book. When extending your environment into Microsoft 365 and SharePoint Online, it's important to continue providing high-quality end user experiences for locating data.

Configuring cloud hybrid search will provide a consistent search experience across on-premises and cloud workloads. In this chapter, we're going to cover the following topics:

- Configuring cloud hybrid search
- Troubleshooting cloud hybrid search

By the end of this chapter, you should have gained an understanding of how to set up cloud hybrid search and how to resolve common configuration issues.

Let's get started!

Overview of Cloud Hybrid Search

Cloud hybrid search uses an on-premises **Search Service Application** (**SSA**) to crawl resources and export this dataset to the SharePoint Online index. Its core benefits, as mentioned in Chapter 10, *Overview of SharePoint Hybrid*, are as follows:

- Users can see unified search results from multiple on-premises sources and cloud resources.
- Cloud hybrid search doesn't require any additional upgrades on-premises as long as the base prerequisites are met.

- It has a lower total cost of ownership for search since no additional on-premises hardware or capacity needs to be deployed moving forward. The enterprise search index is stored in SharePoint Online.
- Microsoft Office Graph applications, such as Delve, can surface content to users.

While cloud hybrid search does have significant benefits, as outlined in the previous list, it also comes with some caveats:

- **Cross-site publishing search**: A cross-site publishing search is not available with a hybrid search.
- **On-premises results for cloud data**: SharePoint Server does not automatically return results for content stored in SharePoint Online. To display SharePoint Online results on your SharePoint on-premises environment, you must configure your on-premises environment to retrieve search results from SharePoint Online.
- **Search verticals**: Search verticals are not migrated or available cross-premises. If you currently use search verticals in your SharePoint Server environment, you must recreate them in SharePoint Online.
- **eDiscovery**: eDiscovery for SharePoint Online is managed in the Security & Compliance Center. SharePoint Online eDiscovery cannot index or search content in SharePoint Server on-premises, resulting in discovery managers having to perform discovery searches in multiple places.
- **Usage reports**: Usage reports are based on information stored in their respective environments. SharePoint Online usage reports do not contain information regarding on-premises user activity or vice versa.
- **Custom search scopes**: Custom search scopes are a legacy SharePoint Server 2010 feature. You'll need to use result sources in SharePoint Online.
- **Best Bets**: **Best Bets** is also a SharePoint Server 2010 feature. You'll need to use result sources in SharePoint Online for this as well.
- **Custom security trimming**: Custom security trimming controls are not supported in SharePoint Online.
- **Multi-tenancy**: SharePoint Online cannot preserve tenant isolation on a multi-tenant SharePoint Server 2013 or SharePoint Server 2016 farm. You'll likely need to transfer data to separate on-premises farms to ensure security boundaries are maintained.
- **Thesaurus**: Thesaurus features are not available on SharePoint Online.
- **The Content Enrichment web service**: The Content Enrichment web service is not available on SharePoint Online.

- **Custom entity extraction**: SharePoint Online does not support custom entity extraction.
- **Index reset for on-premises content**: It is not possible to remove search results for on-premises content if the search service has crawled it. To remove on-premises content from search results, remove the on-premises content source or create an on-premises crawl rule to exclude the content from the search.

It's important to understand these differences and plan for any mitigations or user communication.

Configuring Cloud Hybrid Search

As mentioned in `Chapter 10`, *Overview of SharePoint Hybrid*, there are two supported hybrid search topologies:

- **Cloud hybrid search**: Where users perform searches and are returned results from a single, consolidated index
- **Hybrid federated search**: Where users perform one search and results are returned from both the cloud and on-premises indices separately

Microsoft recommends using cloud hybrid search with the default search configuration. For more information about hybrid federated search, please see `https://docs.microsoft.com/en-us/sharepoint/hybrid/learn-about-hybrid-federated-search-for-sharepoint`.

The following sections will walk you through configuring cloud hybrid search in a SharePoint Server environment.

Prerequisites

Prior to configuring a hybrid search option for SharePoint, you will need to meet the following prerequisites.

For SharePoint Server 2013 only, you will need the following:

- SharePoint Server 2013 Service Pack 1 and the January 2016 public update, available at `https://docs.microsoft.com/en-us/officeupdates/sharepoint-updates`

For SharePoint Server 2013, 2016, and 2019, you will need the following:

- 100 GB storage on the server hosting the search
- The Microsoft 365 or SharePoint Online tenant
- Administrative accounts:
    - A cloud account with a global administrator, application administrator, or cloud application administrator role.
    - The on-premises account must be a member of **Domain Admins**.
    - The on-premises account must be a member of the SharePoint Farm Administrators group.
    - The on-premises account must have the `securityadmin` server role in the farm's SQL server instance.
    - The on-premises account must be a member of the `db_owner` fixed database role on SharePoint databases.
    - The on-premises account must be a member of the local Administrators group on the server where tasks will be performed.
- The Hybrid Configuration Wizard, located at `https://go.microsoft.com/fwlink/?linkid=867176`, or the `CreateCloudSSA.ps1` and `Onboard-CloudHybridSearch.ps1` scripts, found on Microsoft Download Center at `https://www.microsoft.com/en-us/download/details.aspx?id=51490`
- Microsoft Online Services Sign-In Assistant for IT Professionals RTW, located at `https://go.microsoft.com/fwlink/?LinkID=286152`, installed on the search and central administration servers
- The Azure **Active Directory** (**AD**) module for Windows PowerShell, located at `https://docs.microsoft.com/en-us/powershell/azure/active-directory/install-msonlinev1?view=azureadps-1.0`, installed on the search and central administration servers
- The URL of the SharePoint Online root site collection (`https://<tenant>.sharepoint.com`)
- Service accounts:
    - A search service account
    - A managed account for crawling content

The service accounts are normal AD user accounts. They do not require any particular group membership or rights. You will need to register the account you use for crawling content. You can use the `New-SPManagedAccount` cmdlet to register an account:

```
$Credential = Get-Credential -UserName <DOMAIN\cloudsearchmanagedaccount>
New-SPManagedAccount -Credential $Credential
```

You can also opt to deploy additional SharePoint application servers if you want to configure a cloud SSA in a fault-tolerant design. This is not required.

Once these prerequisites have been met, you can move on to preparing Azure AD.

Preparing Azure AD for SharePoint Hybrid

To configure hybrid search services for SharePoint Online, you must subscribe to a Microsoft 365 plan that contains SharePoint Online and connect your on-premises directory to SharePoint Online with Azure AD Connect.

Configuring Azure AD Connect is beyond the scope of this book, but you can learn more about hybrid identity at `https://docs.microsoft.com/en-us/azure/active-directory/hybrid/`.

After the hybrid identity has been established, you can begin configuring the SharePoint cloud hybrid search.

Creating a cloud SSA

There are two methods for configuring cloud hybrid search—using the Hybrid Configuration Wizard or using pre-built PowerShell scripts. Both ways are supported. While the Hybrid Configuration Wizard method is newer and is the method most people use, we'll show both processes.

The terms **Hybrid Picker** and **Hybrid Configuration Wizard** are used interchangeably in both the Microsoft documentation and this book. Hybrid Picker was the original name of the configuration tool. It was later renamed Hybrid Configuration Wizard, but not all references were updated.

The Hybrid Configuration Wizard

Using the Hybrid Configuration Wizard is recommended for most organizations. It will configure the cloud search service with the least amount of administrative effort. The configuration provided by the Hybrid Configuration Wizard was originally a single-server cloud SSA, and if you needed high availability, you would have to use the PowerShell method. However, in the latest version of the Hybrid Configuration Wizard, you can now configure a secondary server.

To complete the cloud SSA setup using the Hybrid Configuration Wizard, follow these steps:

1. Launch **Central Administration** (for newer versions of SharePoint) and navigate to **Office 365**, then select **Launch the Hybrid Configuration Wizard**. Alternatively, you can open a web browser directly to SharePoint's Hybrid Picker (`https://go.microsoft.com/fwlink/?linkid=867176`). Both methods will download the newest version of the Hybrid Configuration Wizard:

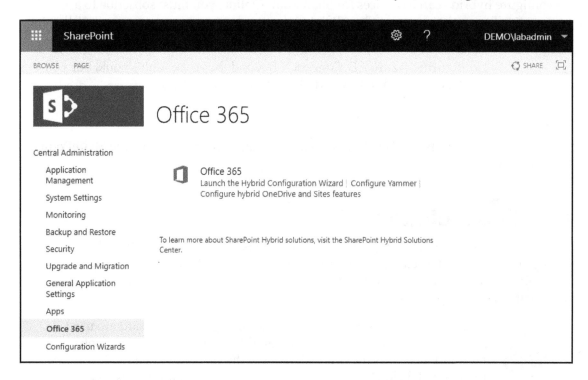

2. Click **Next** to proceed:

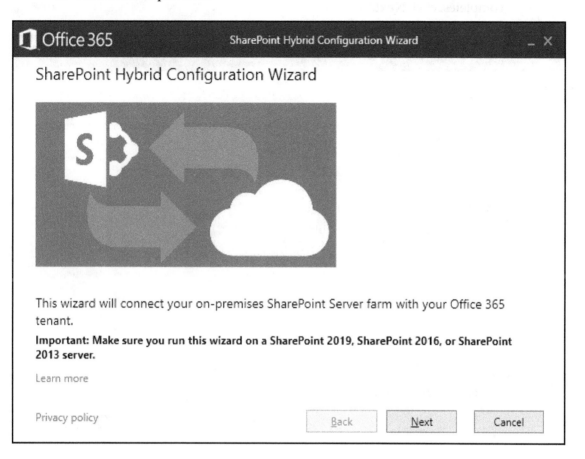

3. Enter the credentials and then click on **Validate credentials**. After the validation completes, click **Next**:

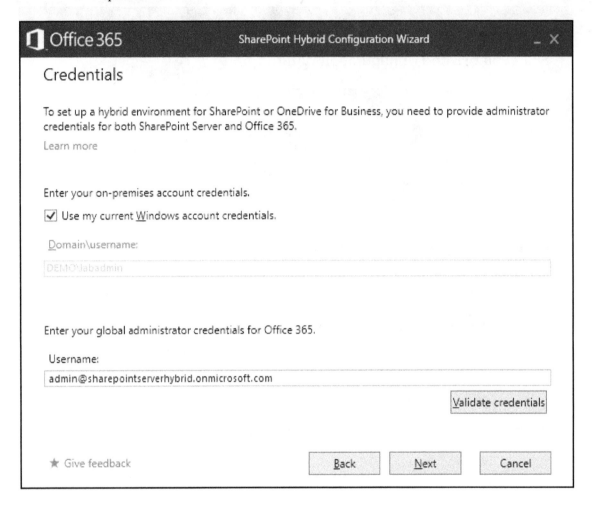

4. Verify that all the prerequisite tests pass. You'll need to resolve any errors before proceeding. Click **Next** when ready:

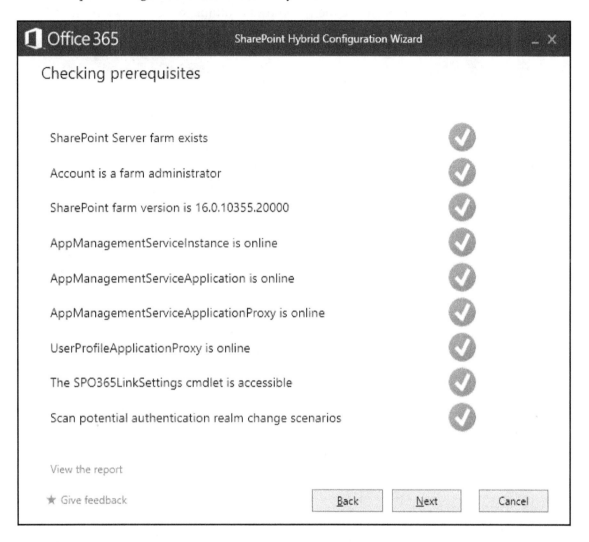

5. If you have configured Workflow Manager (Workflow Manager was discussed in Chapter 3, *Managing and Maintaining a SharePoint Farm*), you'll need to update the security realm to ensure the workflows continue to work. Click **Next**:

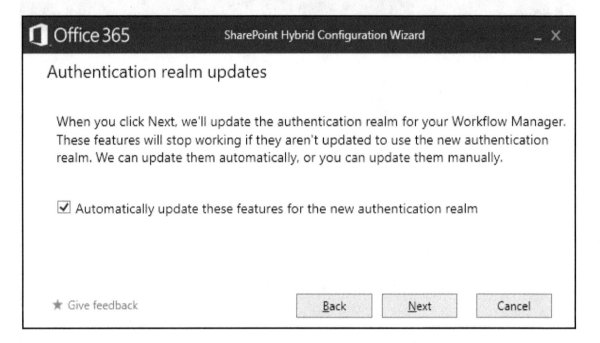

6. Deselect all the preselected options and select **Hybrid Search**:

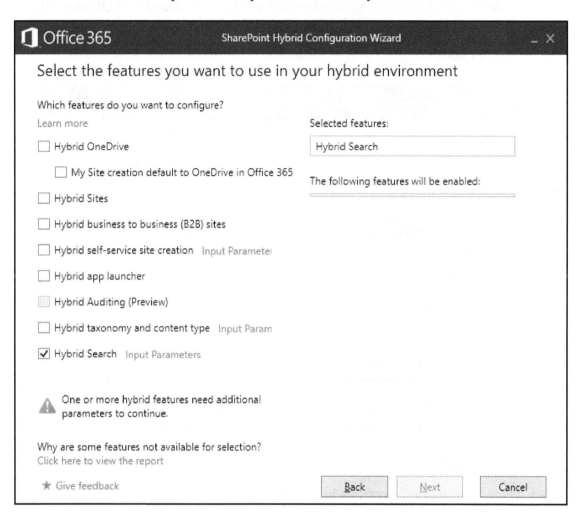

>
>
> Deselecting an option does *not* remove the configuration if it was previously configured.

7. Click on the **Input Parameters** link next to **Hybrid Search**.

8. Enter values for the primary (and, optionally, secondary) servers that will run cloud hybrid search, the SSA name that will be created, and the SQL server that will host the on-premises database. Select the checkbox to confirm you have installed the Azure AD PowerShell module and Sign-In Assistant, and then click **OK**:

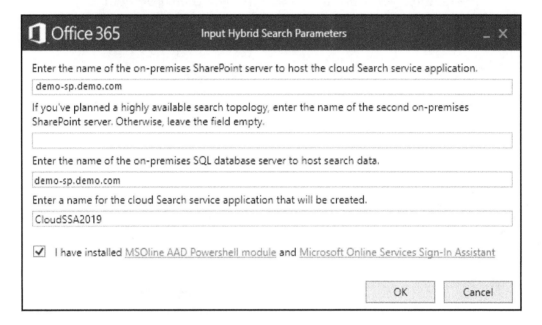

9. Click **Next**. The wizard will configure hybrid search.
10. If any issues are detected during configuration, note and resolve them as indicated.
11. Click **Close**.

Let's see how to configure PowerShell in the next section.

PowerShell

You can also use PowerShell to configure the cloud SSA using the following steps:

1. Launch **SharePoint Management Shell** on a server hosting Central Administration.
2. Navigate to the folder where you have downloaded the PowerShell scripts for configuring the SSA.

3. Run `CreateCloudSSA.ps1` and follow the prompts to provide the necessary values:

- **SearchServerName**: The primary server that will run the cloud SSA.
- **SearchServerName2**: If you are deploying in a fault-tolerant design, this will be the second server running the cloud SSA.
- **SearchServiceAccount**: The service account (in the `DOMAIN\username` format) for the search service.
- **SearchServiceAppName**: The name for the cloud SSA that will be created.
- **DatabaseServerName**: The name of the SQL server to be used for the SSA.

After entering these details, the script will configure the cloud SSA. Once the cloud SSA has been created, you can proceed with connecting it to SharePoint Online.

Connecting the cloud SSA to SharePoint Online

If you used the Hybrid Configuration Wizard to configure a hybrid search, the following steps are not necessary (they're already performed in the wizard). However, if you ran the configuration from PowerShell, you'll need to complete it by connecting the cloud search service to SharePoint Online:

1. Launch **SharePoint Management Shell** and switch to the directory where the cloud hybrid search configuration scripts have been downloaded.
2. From **SharePoint Management Shell**, run the following command. Enter your SharePoint Online admin or global admin credentials when prompted:

   ```
   $Credential = Get-Credential
   ```

3. Run the following command using your organization's SharePoint Online URL and the name of the cloud SSA used when running the `CreateCloudSSA.ps1` script:

   ```
   .\Onboard-CloudHybridSearch.ps1 -CloudSsaId <CloudSsaID> -
   PortalUrl https://<tenant>.sharepoint.com -Credential $Credential
   ```

When running the onboarding script, be sure to check for any items that fail. An example of running the onboarding script is shown in the following screenshot:

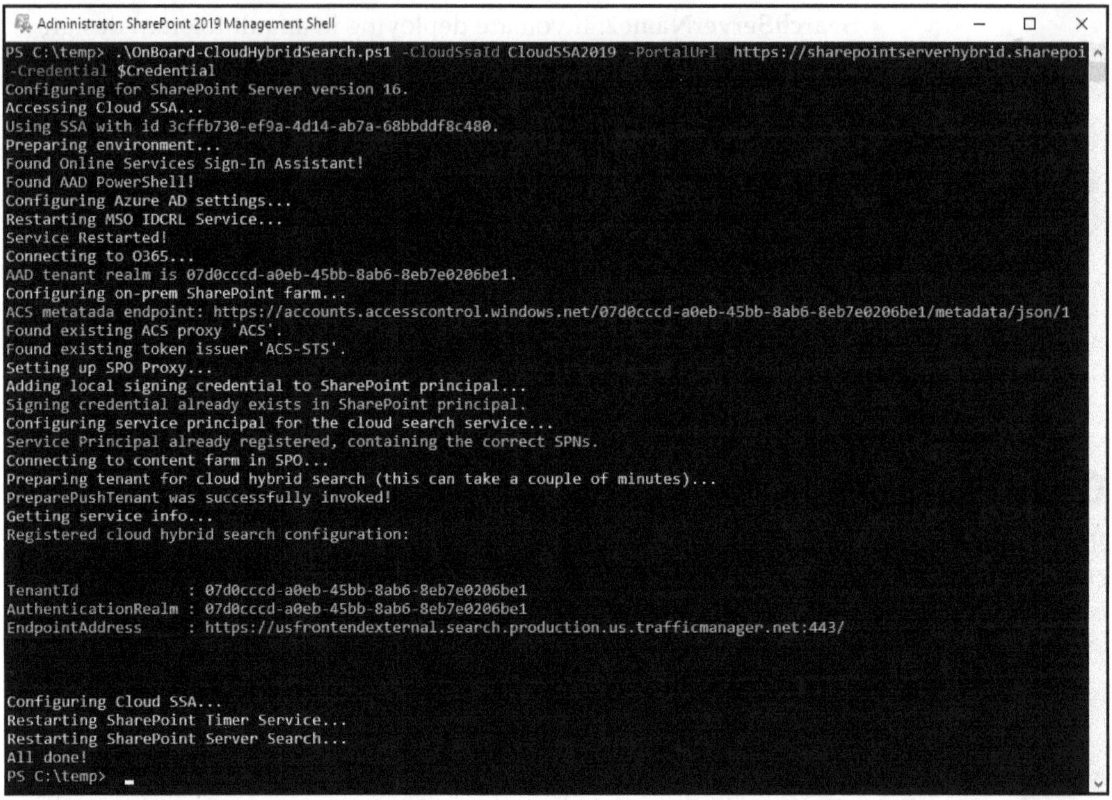

```
Administrator: SharePoint 2019 Management Shell                                          —    □    ×
PS C:\temp> .\OnBoard-CloudHybridSearch.ps1 -CloudSsaId CloudSSA2019 -PortalUrl https://sharepointserverhybrid.sharepoi
 -Credential $Credential
Configuring for SharePoint Server version 16.
Accessing Cloud SSA...
Using SSA with id 3cffb730-ef9a-4d14-ab7a-68bbddf8c480.
Preparing environment...
Found Online Services Sign-In Assistant!
Found AAD PowerShell!
Configuring Azure AD settings...
Restarting MSO IDCRL Service...
Service Restarted!
Connecting to O365...
AAD tenant realm is 07d0cccd-a0eb-45bb-8ab6-8eb7e0206be1.
Configuring on-prem SharePoint farm...
ACS metadata endpoint: https://accounts.accesscontrol.windows.net/07d0cccd-a0eb-45bb-8ab6-8eb7e0206be1/metadata/json/1
Found existing ACS proxy 'ACS'.
Found existing token issuer 'ACS-STS'.
Setting up SPO Proxy...
Adding local signing credential to SharePoint principal...
Signing credential already exists in SharePoint principal.
Configuring service principal for the cloud search service...
Service Principal already registered, containing the correct SPNs.
Connecting to content farm in SPO...
Preparing tenant for cloud hybrid search (this can take a couple of minutes)...
PreparePushTenant was successfully invoked!
Getting service info...
Registered cloud hybrid search configuration:


TenantId            : 07d0cccd-a0eb-45bb-8ab6-8eb7e0206be1
AuthenticationRealm : 07d0cccd-a0eb-45bb-8ab6-8eb7e0206be1
EndpointAddress     : https://usfrontendexternal.search.production.us.trafficmanager.net:443/


Configuring Cloud SSA...
Restarting SharePoint Timer Service...
Restarting SharePoint Server Search...
All done!
PS C:\temp> _
```

Next, we'll create a content source for the cloud hybrid search to index.

Creating a content source for the cloud hybrid search

After the cloud SSA has been created and connected to SharePoint Online, you must create a content source to be incorporated into the SharePoint Online search index. This content source can be any source supported by the SharePoint server, including file shares, websites, SharePoint sites, Exchange public folders, or other custom data sources.

To add a content source, follow these steps:

1. Launch **Central Administration**. Select **Application Management**, then select **Manage service applications**.
2. On the **Manage service applications** page, select the cloud SSA that you previously created and onboarded.
3. Under **Crawling**, select **Content Sources**:

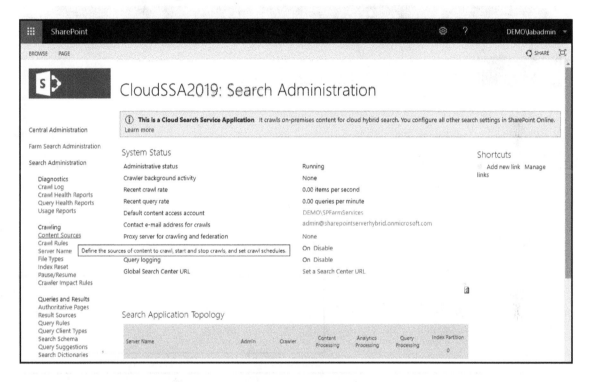

4. Click on **New Content Source** to create a selection of content to crawl.
5. Under **Name**, type in a name for the content source.
6. Under **Content Source Type**, select the type of content that will be crawled from the available options.

7. Under **Start Address**, enter the addresses that will be included in the content search. If you are searching SharePoint sites or websites, for example, you can type in `http://server`. If you intend to crawl file shares, type in the addresses as `\\server\share`:

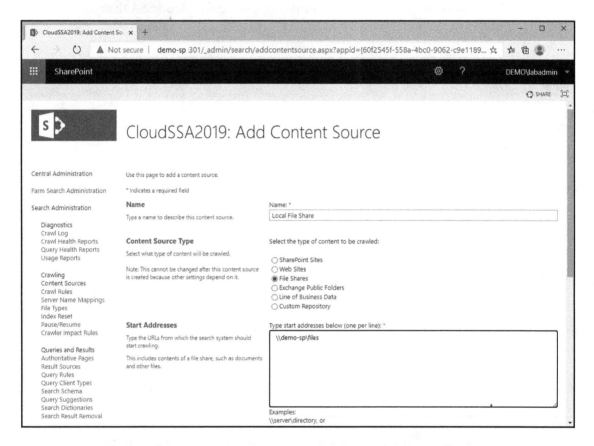

8. Under **Crawl Settings**, select the behavior for crawling—either **Folder and all subfolders of each start address** (default) or just **Folder of each start address** (top-level folders only).

9. Under **Crawl Schedules**, configure a schedule for full and incremental crawls (if desired).

10. Under **Content Source Priority**, configure whether this content source will have a **High** or **Normal** priority. Selecting the **High** priority option prioritizes this content source's processing over content sources set with the **Normal** priority option.

11. Click **OK** to save.

12. Click on the content source and select **Start Full Crawl**, then click **OK** to confirm:

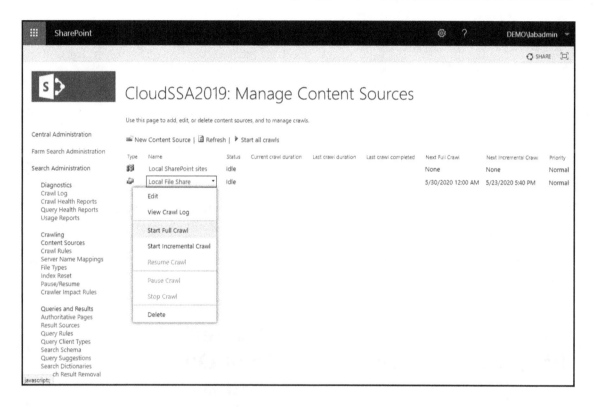

After the content source crawls are completed, navigate to SharePoint Online and perform a search for `IsExternalContent:true`:

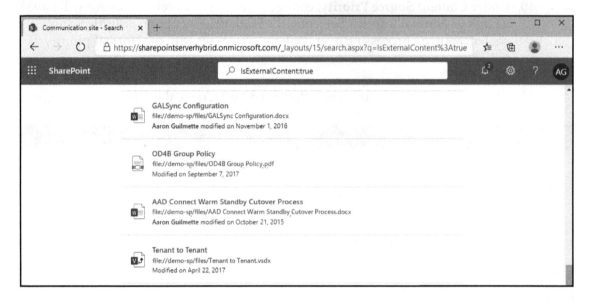

The `IsExternalContent:true` property search shows content that is external to where the search originated (in this case, content external to SharePoint Online). The previous search was executed in SharePoint Online, and the dataset shows results from the local SharePoint environment (`file://server/share`), confirming that cloud hybrid search is working correctly.

Configuring the SharePoint Server search to display results from SharePoint Online

You can also configure the on-premises SharePoint Server environment to display results from SharePoint Online. This way, your users get the same results from either environment.

To configure on-premises search results to include SharePoint Online sources, follow these steps:

1. Launch **Central Administration**.
2. Select **Application Management** and then click on **Manage service applications**.
3. Select the cloud SSA.
4. Under **Queries and Results**, select **Result Sources**.
5. Select **New Result Source**:

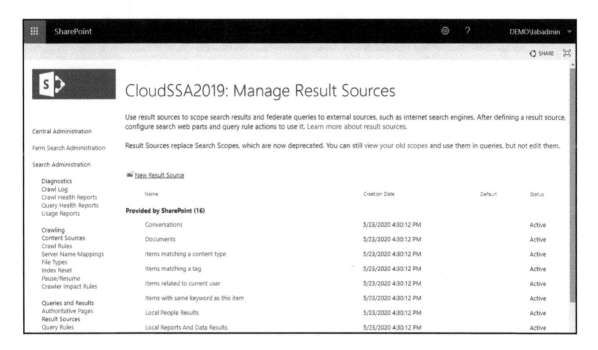

6. Under **General Information**, enter a name for the result source, such as `SharePoint Online`:

- Under **Protocol**, select the **Remote SharePoint** radio button.
- Under **Remote Service URL**, enter the top-level URL of your SharePoint Online tenant (`https://<tenant>.sharepoint.com`):

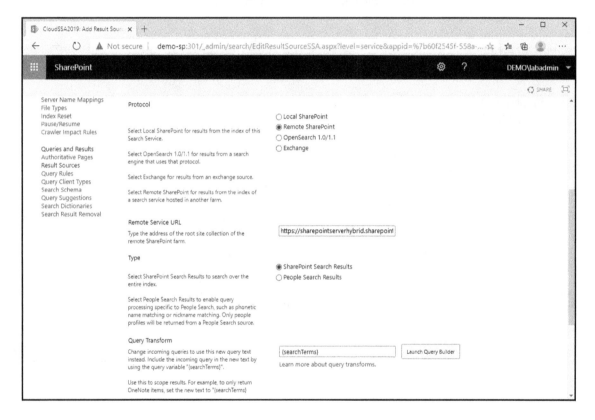

7. Under **Type**, ensure that the **SharePoint Search Results** radio button is selected.
8. Under **Query Transform**, leave the default value (`{searchTerms}`).
9. Under **Credentials Information**, ensure that the **Default Authentication** radio button is selected.
10. Click **Save**.

11. Select the down arrow next to the newly created result source and then select **Set as Default**:

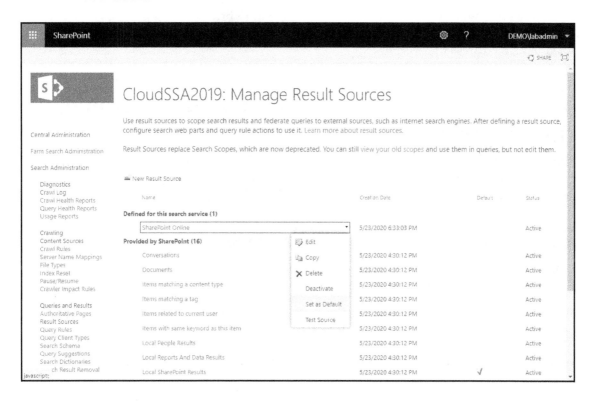

Log on to the SharePoint Server search site with an identity that is federated and licensed in the Microsoft 365 tenant for SharePoint Online. Using the same method as when we limited SharePoint Online to only display on-premises search results, perform a search using the `IsExternalContent:true` search property.

Next, we'll look at how to address troubleshooting.

Troubleshooting Cloud Hybrid Search

There are times when configuring or managing cloud hybrid search doesn't go as expected. Most issues can be easily fixed. The following common troubleshooting scenarios can be used to help address these issues.

The cloud hybrid search setup fails to complete due to networking

Cloud hybrid search requires access to SharePoint Online. If your environment is behind a proxy server or another network-filtering appliance, you should try to bypass the proxy server or appliances to reach SharePoint Online. If bypassing the proxy is not possible, you may need to configure the WinHTTP proxy settings via `netsh`.

You can use `netsh` to import Internet Explorer's proxy configuration into a WinHTTP proxy configuration:

```
netsh winhttp import proxy source=ie
```

For more information on the uses and capabilities of `netsh`, see `https://docs.microsoft.com/en-us/windows-server/networking/technologies/netsh/netsh-contexts`.

Once you've updated the proxy settings, rerun the Hybrid Configuration Wizard.

If the Hybrid Configuration Wizard fails

The Hybrid Configuration Wizard may fail with an unspecified error. The log file, located in `%appdata%\Microsoft\SharePoint Hybrid Configuration`, can reveal additional details, such as pre-existing services or access problems:

```
20200523_162131.log - Notepad                                                          —    □    ×
File  Edit  Format  View  Help
2020.05.23 16:21:46.141 WARNING [Activity=OnPremises Connection Validation] The remote server returned an error: (401) Unauthorized.
2020.05.23 16:21:46.141         [Activity=OnPremises Connection Validation] Connecting to http://demo-sp:5985/wsman...
2020.05.23 16:21:46.157 WARNING [Activity=Tenant Connection Validation] The remote server returned an error: (401) Unauthorized.
2020.05.23 16:21:46.157         [Activity=Tenant Connection Validation] Validating tenant credential...
2020.05.23 16:21:46.172 WARNING [Activity=Tenant Connection Validation] Wrong tenant credential.
2020.05.23 16:21:46.172         [Activity=Tenant Connection Validation] Wrong tenant credential.
2020.05.23 16:21:46.172         [Activity=Tenant Connection Validation] FINISH Time=5499.9ms
2020.05.23 16:21:46.657         [Activity=OnPremises Connection Validation, Provider=SharePointPowerShellProvider] Adding PSSnapIn Microsoft.ShareP
2020.05.23 16:21:48.469         [Activity=OnPremises Connection Validation, Provider=SharePointPowerShellProvider] Opening Runspace.
2020.05.23 16:21:49.594         [Activity=OnPremises Connection Validation] Connected in 3448ms
2020.05.23 16:21:49.610         [Activity=OnPremises Connection Validation] Getting configuration data...
2020.05.23 16:21:49.610         [Activity=OnPremises Connection Validation] Data retrieved
2020.05.23 16:21:49.610         [Activity=OnPremises Connection Validation] FINISH Time=8937.4ms
2020.05.23 16:21:49.626         [Activity=OnPremises Connection Validation, Provider=SharePointPowerShellProvider] Disposing Runspace.
2020.05.23 16:22:01.302         [Activity=OnPremises Connection Validation] START
```

In the preceding screenshot, the credentials were marked as **user must change password on the next login**, which prevented them from actually logging in. For any errors during configuration, look at the log first to determine the error details.

The cloud hybrid search fails to return expected results

There can be a number of reasons for this, but the most common reason is that the data source you're indexing on-premises (sites or files shares) only has permissions granted to the **Domain Users** group. Azure AD Connect does not synchronize objects with the IsCriticalSystemObject attribute set to True. **Domain Users**, among other built-in groups, has that attribute set to True, blocking it from synchronization. Since SharePoint search results are security-trimmed, you cannot see the indexed results.

Check the content sources you're indexing—if **Domain Users** is the only security principal (user or group) that is granted access, update the access control list on the object to a user or group that is synchronized to SharePoint Online and then try again after the search index has updated.

Additionally, the crawl account may not have sufficient permissions to index the files, sites, or other resources. If that is the case, then you will need to grant the account access and restart the crawl. The crawl account requires the **Manage auditing and security log** privilege. You can review the crawl log by selecting **Crawl Log** under the **Diagnostics** section of the cloud SSA:

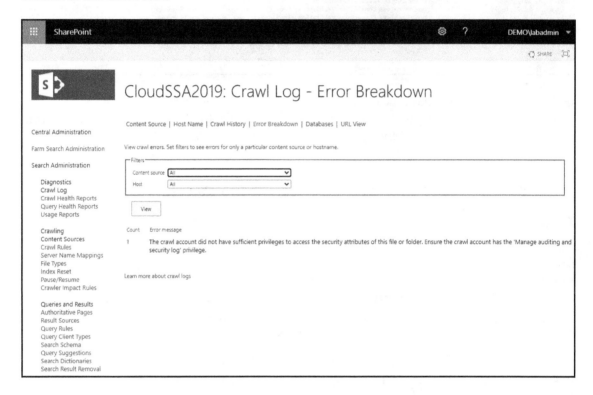

If any of the preceding methods don't allow you to resolve the cloud hybrid search issues, you'll need to open a ticket with Microsoft Support.

Summary

In this chapter, we covered configuring SharePoint hybrid search using both the Hybrid Configuration Wizard and manually through PowerShell. SharePoint hybrid search allows you to configure both the SharePoint Server and SharePoint Online environments to return a complete list of results to your users, regardless of where they initiate the query. This is critical in ensuring users can locate the data they need. Configuring hybrid search through either PowerShell or the Hybrid Configuration Wizard will give organizations the ability to work more seamlessly between their cloud and on-premises platforms.

In the next chapter, we'll continue working with cross-premises scenarios by introducing the data gateway.

14
Implementing a Data Gateway

In the previous few chapters, you learned about configuring various SharePoint Hybrid services, such as search. We're going to continue building on those concepts with data gateways. A data gateway, in Microsoft terminology, is a service that can be used to provide cloud-based Power Platform and Azure services with access to on-premises data sources, such as SharePoint Server and SQL Server.

A data gateway allows organizations to keep services or applications in their own managed infrastructure while providing access to that data via a secure connection. It acts in a similar way to a reverse proxy, providing secure access to on-premises resources. In the following diagram, you can see how the data gateway works between the Power Platform services and dashboards in Microsoft 365 and on-premises data sources:

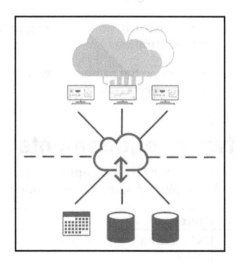

In this chapter, we'll focus on the following areas:

- Planning the implementation of an on-premises data gateway
- Installing and configuring an on-premises data gateway
- Managing an on-premises data gateway
- Troubleshooting common issues

Let's get started!

Planning the implementation of an on-premises data gateway

Microsoft allows organizations to configure data gateways in two modes:

- **Standard mode**: This is the default data gateway configuration mode. It allows multiple users to connect to multiple data sources.
- **Personal mode**: This data gateway configuration mode allows just a single user to connect to data sources. The configuration cannot be shared with others. This mode can only be used for Power BI.

For the purposes of administering infrastructure, we're going to focus on the standard mode on-premises data gateway configuration.

In the following sections, we will examine the prerequisites and requirements for the successful installation and configuration of a data gateway.

Server and software requirements

In order to configure a data gateway, the target computer must meet certain minimum requirements. Microsoft has also published some recommended requirements:

Configuration type	Notes
Minimum requirements	.NET Framework 4.6 (gateway release August 2019 and earlier) .NET Framework 4.7.2 (gateway release September 2019 and later) A 64-bit version of Windows 8 or a 64-bit version of Windows Server 2012 R2

Recommended configuration	An 8-core CPU 8 GB of memory A 64-bit version of Windows Server 2012 R2 or later **Solid state drive** (**SSD**) storage for spooling

The Microsoft documentation does not currently list a specific minimum memory requirement for a data gateway installation. The minimum system memory requirement for Windows 8 64-bit edition is 2 GB. The minimum system memory requirement for Windows Server 2012 R2 is 512 MB.

In addition to the preceding requirements table, there are some other considerations for installing the gateway:

- Gateways are not supported on Windows Server Core installations.
- Gateways cannot be installed on **Active Directory** (**AD**) domain controllers.
- Gateways that use Windows authentication should be installed on a computer that is a member of the same AD domain as the data sources it will be used to access.
- A computer can only have one standard gateway configured on it.

It's not recommended to install a gateway on computers with intermittent accessibility (such as laptops or computers that can be disconnected from the internet) or computers that use wireless network adapters to connect to the internet.

Next, we'll look at the networking requirements.

Networking requirements

The purpose of the data gateway is to provide a conduit from your on-premises data to services in Office 365, such as Power Automate and Power BI. You must plan for and ensure proper communication and connectivity in order to utilize a data gateway. The networking requirements can be divided into two parts: endpoints and proxy servers. We'll address each of these now.

Endpoints

Endpoints refer to any data source or service that you're connecting to. The computer(s) hosting data gateways must be able to communicate with on-premises data sources over the appropriate ports. Additionally, the computer(s) hosting the data gateway must be able to communicate with the following Microsoft endpoints:

Domain names	Outbound ports	Description
`*.download.microsoft.com`	80	Used to download the installer. The gateway app also uses this domain to check the version and gateway region.
`*.powerbi.com`	443	Used to identify the relevant Power BI cluster.
`*.analysis.windows.net`	443	Used to identify the relevant Power BI cluster.
`*.login.windows.net`, `login.live.com`, and `aadcdn.msauth.net`	443	Used to authenticate the gateway app for Azure AD and OAuth 2.
`*.servicebus.windows.net`	5671–5672	Used for **Advanced Message Queuing Protocol (AMQP)**.
`*.servicebus.windows.net`	443 and 9350–9354	Listens on the service bus relay over TCP. Port 443 is required to get Azure access control tokens.
`*.core.windows.net`	443	Used by dataflows to write data to Azure Data Lake.
`login.microsoftonline.com`	443	Used to authenticate the gateway app.
`*.msftncsi.com`	443	Used to test internet connectivity if the Power BI service can't reach the gateway.
`*.microsoftonline-p.com`	443	Used to authenticate the gateway app for Azure AD and OAuth 2.
`dc.services.visualstudio.com`	443	Used by App Insights to collect telemetry.

The data gateway requires *outbound connectivity only*. No inbound firewall rules need to be configured.

Proxy servers

A lot of organizations require intermediary devices, such as proxy servers, to control access to internet services. If your organization requires a proxy server to connect to the internet, the best practice is to request an exception for the server(s) hosting the data gateway to be excluded from proxy configuration.

However, this may not be possible. In the event that you cannot bypass the proxy for connections to the internet, you may need to request credentials to be used by the gateway service.

As previously noted, data gateways are not recommended for installation on computers with wireless internet connectivity.

Security and credentials

In order to install and configure the gateway, you will need to be a local administrator on the computer where the software will be configured. Additionally, standard gateways require administrative access (such as the Power Platform service admin or global admin roles) to the Power Platform data gateway configuration pages.

The data gateway service is configured with a local account called NT SERVICE\PBIEgwService. You can change this to a domain user account or a **Managed Service Account (MSA)**. If your organization requires periodic password changes as part of its security policy, you may want to configure the data gateway service to use an MSA.

High-availability requirements

Data gateways can be configured in a cluster to allow fault tolerance when accessing on-premises data sources. Since a computer hosting a data gateway can only have a single standard gateway configured on it, you'll need a second computer if you want to configure a cluster.

Recovery key

The primary purpose of a recovery key is to encrypt credentials for use with Office 365. The recovery key is necessary when you need to take over, restore, or move the gateway. In older versions of the gateway software, it could not be changed. Newer versions, starting with November 2019 (3000.14.39), support changing the recovery key. To change the recovery key, you will need access to the previous recovery key.

All recovery keys should be recorded for safekeeping in the event that it is needed for maintenance or disaster-recovery operations.

Now that we've gone over the requirements, we can proceed with installing a data gateway.

Installing and configuring an on-premises data gateway

In order to start accessing and visualizing your on-premises data, you'll need to install a gateway. Installing and configuring a gateway involves several steps:

- Configuring the networking
- Downloading and installing the software

Use the following process to download, install, and configure the data gateway software.

Configuring networking

There are two sets of networking requirements—a data gateway to the internet and a data gateway to on-premises data sources. You'll need to configure both.

The previous section detailed the outbound network connectivity requirements. You'll need to work through your organizational change-control process to allow connectivity from the computer(s) hosting the data gateway software to the required endpoints. To access on-premises datasets, you'll need to ensure the data gateway can communicate with on-premises application and database servers. Making these configuration changes is beyond the scope of this book, but it's important to know that your organization may require changes in order to successfully deploy and configure a data gateway.

If you need to configure additional settings (such as a proxy server), you will need to edit the configuration after the data gateway has been installed. Use the following sections to update the configuration after installation.

Configuring a proxy server

The proxy server configuration is maintained in two files outside the data gateway desktop application. The files that may need to be modified are `C:\Program Files\On-premises data gateway\enterprisegatewayconfigurator.exe.config` and `C:\Program Files\On-premises data gateway\Microsoft.PowerBI.EnterpriseGateway.exe.config`.

If manual proxy configurations are required, both files must be edited to contain the same data. If you have updated other .NET application configuration files before, the format will be familiar:

```
<system.net>
    <defaultProxy useDefaultCredentials="true" />
</system.net>
```

A common update may be to configure specific proxy server settings. In the following example, a proxy server of `10.0.0.80` is used with port `8080`:

```
<system.net>
    <defaultProxy useDefaultCredentials="true">
        <proxy
            autoDetect="false"
            proxyaddress="http://10.0.0.80:8080"
            bypassonlocal="true"
            usesystemdefault="true"
        />
    </defaultProxy>
</system.net>
```

After updating the configuration, you may need to restart the data gateway service.

Configuring proxy server authentication

If your proxy requires Windows authentication, it is recommended that you configure the data gateway application to use a domain account.

After updating the configuration, you may need to restart the data gateway service.

Next, we'll review the steps for installing the data gateway software.

Downloading and installing the software

Once the server and software networking prerequisites have been met, you can proceed with downloading and installing the data gateway application:

1. On the computer where the data gateway will be installed, open a web browser and navigate to `https://go.microsoft.com/fwlink/?LinkId=2116849`.
2. When prompted, save the file.
3. After the download is complete, open the file and run it.
4. Accept the default location installation (per the Microsoft documentation). Select the checkbox to agree to the terms of use, and then click **Install**.
5. Once the initial software installation is complete, use administrative credentials to sign in to the Office 365 tenant.
6. Select the **Register a new gateway on this computer** radio button and click **Next**:

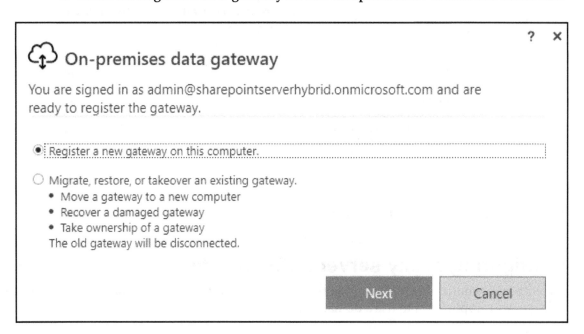

7. Create a name for your data gateway, enter a recovery key to be used in the future (for recovery operations or for creating a gateway cluster), select a region (if desired), and click **Configure**:

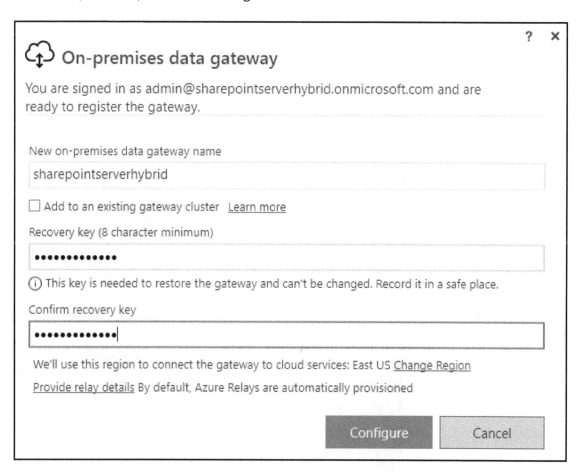

 The **Change Region** option is only available for public commercial cloud customers. If your Office 365 tenant is in a sovereign cloud, such as Office 365 Government Community Cloud, you must use the region associated with your tenant.

8. Review the app notifications, if any. Click **Close** to close the **On-premises data gateway** app:

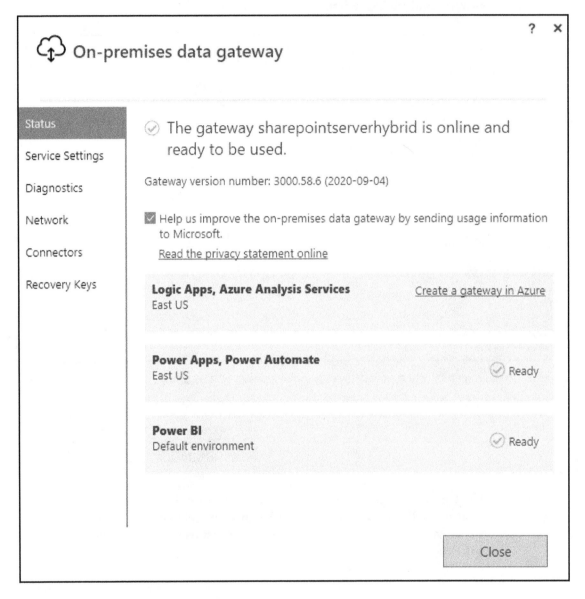

The data gateway has been installed and is ready to use.

Creating a cluster

If you have determined that your server has high-availability needs for the data gateway, you can configure a cluster. To add an additional node to create a gateway cluster, take the following steps:

1. On the computer where the data gateway will be installed, open a web browser and navigate to `https://go.microsoft.com/fwlink/?LinkId=2116849`.
2. When prompted, save the file.
3. After the download is complete, open the file and run it.
4. Accept the default location installation (per the Microsoft documentation). Select the checkbox to agree to the terms of use, and then click **Install**.
5. Once the initial software installation is complete, use administrative credentials to sign in to the Office 365 tenant.
6. Select the **Register a new gateway on this computer** radio button and click **Next**.
7. Select a name for the gateway and select the **Add to an existing gateway cluster** checkbox. In the **Available gateway clusters** drop-down list, select the primary node or primary gateway (the first gateway you installed).
8. In the **Recovery key** box, enter the value used for the recovery key when creating the gateway.
9. Click **Configure**.

The data gateway cluster has been installed and is ready to use.

Installation, however, is only the first part of the process. Once the gateway has been installed, you will need to manage the configuration to add data sources and make them available to Office 365. In the next section, we'll review common configuration tasks for data gateways.

Managing an on-premises data gateway

Now that a data gateway has been installed, you can configure data sources to be used with Office 365 Power Platform applications, as well as update the settings for the gateway. In this section, you'll learn how to do the following tasks:

- Add a data source.
- Add a gateway admin.

- Change the gateway service account.
- Change the recovery key.
- Monitor a data gateway.

Use the procedures in the following sections to perform maintenance and management tasks for the data gateway.

Adding a data source

In order to make data available to Office 365 Power Platform services, you'll need to configure one or more data sources. To manage data sources, follow these steps:

1. Log in to the Power BI admin center (`https://app.powerbi.com/admin-portal`).
2. Select the gear icon, and then select **Manage gateways**:

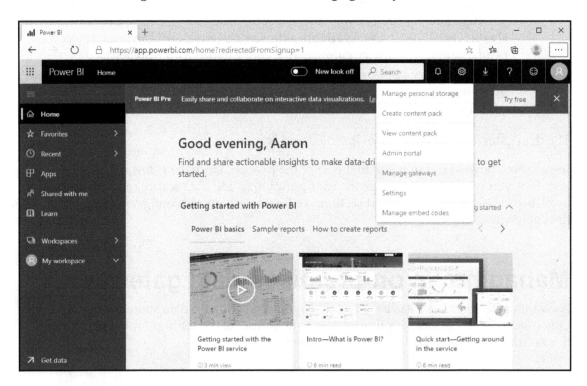

3. Select the gateway to which you want to add a data source and then click on
   the **Add data sources to use the gateway** link:

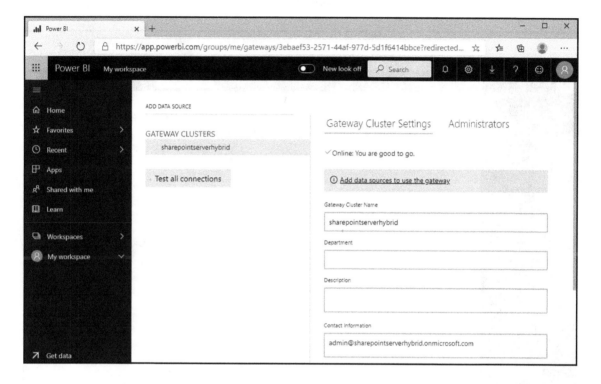

4. Enter a name into the **Data Source Name** field. Select a type and, depending on the data source type, add any additional information, such as a resource URL, an authentication method, or credentials, and click **Add**:

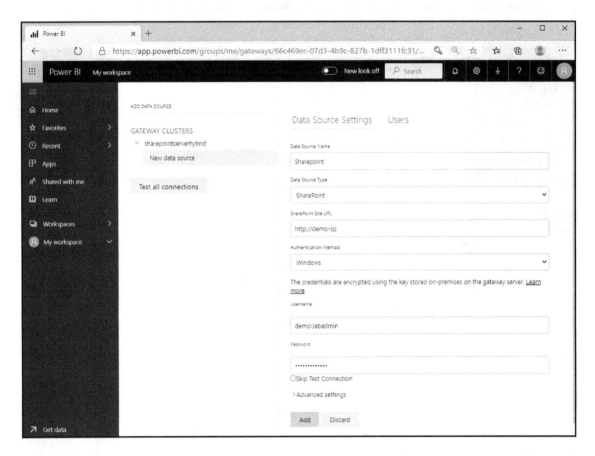

5. Once the connection is successful, select the **Users** tab to configure any additional users for this data source:

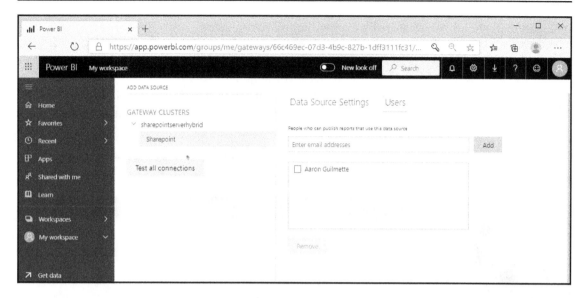

Any data sources that you add through this interface will be available throughout the Power Platform ecosystem, including Power Automate, Power Apps, and Power BI.

It's important to remember when configuring a data source to use the *internal names and addresses* of servers or sites. Since the data gateway resides on your organization's *internal network,* it will use the internal name resolution capabilities of the server where it is installed.

Adding a gateway admin

By default, if you install and configure a gateway, you are made an administrator of the gateway. There may be, however, a need to delegate or add additional individuals to administer the gateway. The option to manage gateways will not be available unless you have either installed a gateway or been added as an administrator of an existing gateway.

You can manage administrators by taking the following steps:

1. Log in to the Power BI admin center (`https://app.powerbi.com/admin-portal`).
2. Select the gear icon, and then select **Manage gateways**.
3. Select the gateway to manage and click on the **Administrators** tab. Start entering the name or address of the user you wish to add as an administrator and click **Add**:

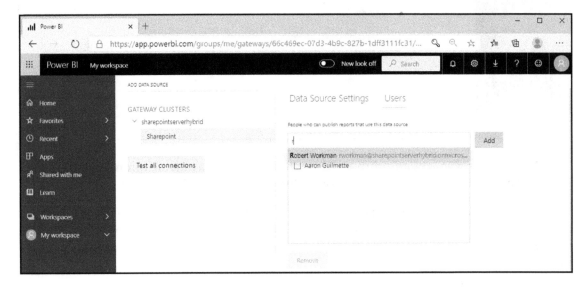

4. You can also remove administrators by selecting the checkbox next to the user and clicking **Remove**.

The selected user can now administer the gateway.

Changing the gateway service account

If you need to change the data gateway service account (for example, to resolve proxy authentication errors or to implement an MSA), you can use the **On-premises data gateway** application to do so.

You can configure the gateway to use either a normal user account or an MSA. You'll see how to do both in the following sections.

Updating the data gateway service to use a group MSA

You might use an MSA if your organization has specific security requirements (such as changing passwords at intervals). MSAs help achieve these organizational roles by allowing the system to maintain and automatically roll over the password, much like how domain computer account passwords are updated.

In this example, we'll configure a new group MSA and then configure the data gateway to use it.

Creating a group MSA

Before you can configure the data gateway to use an MSA, you need to create the account. If you will be configuring a data gateway cluster, you may want to consider using a group MSA (as opposed to a standard MSA) so that multiple computers can use the same service account.

In order to use a group MSA, your forest must be updated to at least the Windows Server 2012 schema. You can check the current version of the schema with the following PowerShell command:

```
(Get-ADObject (Get-ADRootDSE).schemaNamingContext -
properties objectVersion).objectVersion
```

If the value is less than 52, you will need to update the schema using Adprep.exe from Windows Server 2012 or later.

To configure a group MSA, follow these steps:

1. On the computer hosting the data gateway, log in as an administrator with privileges to administer the domain.

In order to create a group MSA account, you must be a member of **Domain Admins** or have been granted rights to create group MSAs. You can carry out all of the account preparation steps on the domain controller, except the step to install the service account on the computer hosting the data gateway.

2. Launch an elevated PowerShell console session.

3. Run the following cmdlet to install the AD **Remote Server Administration Tools** (**RSATs**) if they are not already present:

```
Install-WindowsFeature RSAT-ADDS
```

4. Run the following cmdlet to configure a KDS root key (if one is not already configured):

```
Add-KdsRootKey -EffectiveTime ((Get-Date).AddHours(-10))
```

5. Run the following command to create the group MSA. In this example, the name I'm going to use for the service account is demodg-svc and the computers hosting the data gateway where this service account will be used are demo-sp and demo-sp14. For the DnsHostName parameter, I simply used the name of the service account and appended the domain suffix.

When specifying the computers to be allowed to use the group MSA in the PrincipalsAllowedToRetrieveManagedPassword value, you'll need to append a dollar sign ($) character to the end of the computer hostname, or specify the computer as CN=computername,OU=organizationalUnit,dc=domain,dc=com to allow the cmdlet to resolve the computer hostname. Appending a dollar sign *is not necessary* when adding additional target computers using Add-ADComputerServiceAccount.

Typically, when creating group MSAs, you don't need to configure a password. As of the time of writing, the data gateway service app configuration does not support configuring a group MSA natively from the app, so we'll go through a few extra steps to do so later:

```
New-ADServiceAccount -Name "demodg-svc" -
PrincipalsAllowedToRetrieveManagedPassword demo-sp$,demo-sp14$    -
DnsHostName demodg-svc.demo.com -Enabled $True
```

Next, you'll need to add the service account to the computer hosting the data gateway:

This step must be performed on each computer hosting the data gateway. The Install-ADServiceAccount cmdlet requires RSATs to be installed on the server running the cmdlet.

```
Install-ADServiceAccount -Identity demodg-svc
```

Now, you've created and installed the group MSA on the computer hosting the data gateway. In the next section, we'll configure the data gateway to use a new account.

Configuring the service

You can take these steps to change the service account used by the on-premises data gateway and configure the service to use a group MSA:

1. On the computer hosting the data gateway, launch the **Services** applet (`services.msc`).

2. Locate the **On-premises data gateway service** option and double-click on it:

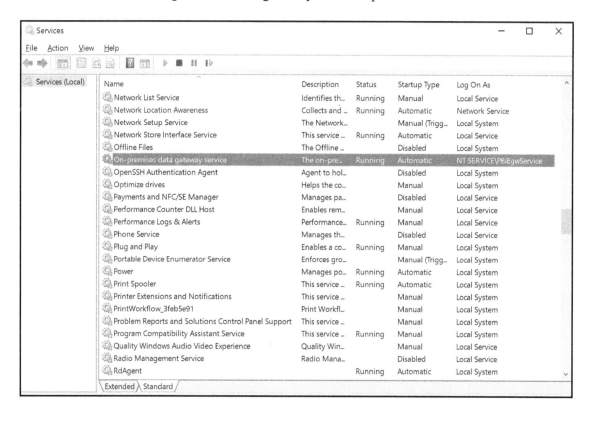

3. Update the service name to the domain user account for the group MSA that you wish to use, and then click **OK**:

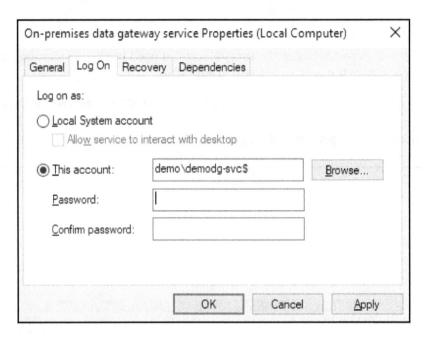

When using an MSA, remember to append a dollar sign ($) to the account name and leave the password value blank.

4. When notified that the **Log On As A Service** right has been granted, click **OK**.
5. Click **OK** again to acknowledge that the service has to be stopped and restarted manually.
6. Restart the service.
7. Launch the **On-premises data gateway** app. When prompted, sign in as an administrator of this data gateway:

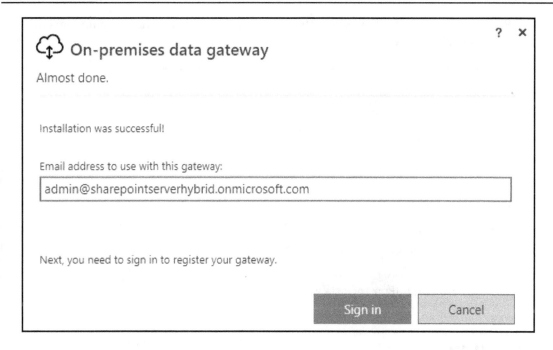

8. Select the **Migrate, restore, or takeover an existing gateway** option and click **Next**:

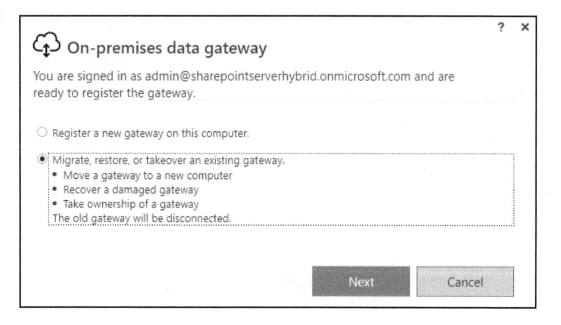

9. Enter the recovery key, select the correct instance (if you have a gateway cluster), and click **Configure**:

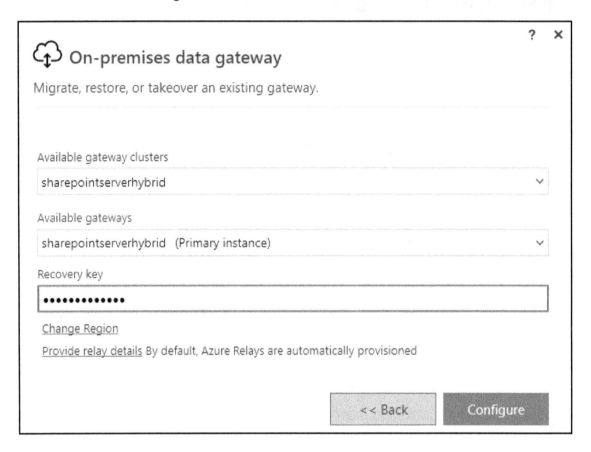

10. Wait as the gateway is recovered.
11. Confirm that the process has completed successfully. Click **Close**.

12. If running a gateway cluster, repeat the process on the additional nodes, selecting the corresponding gateway node/instance, as shown in the following screenshot:

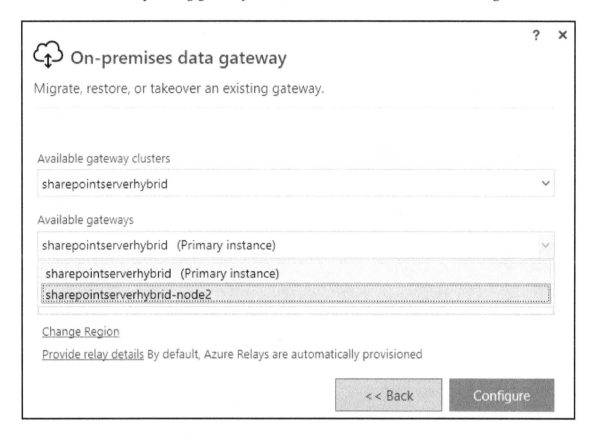

The update process is complete.

Updating the data gateway service to use a domain account

You can take the following steps to change the service account used by the on-premises data gateway and to configure the service to use a normal user account:

1. On the computer hosting the data gateway, launch the **On-premises data gateway** app and select **Service Settings**:

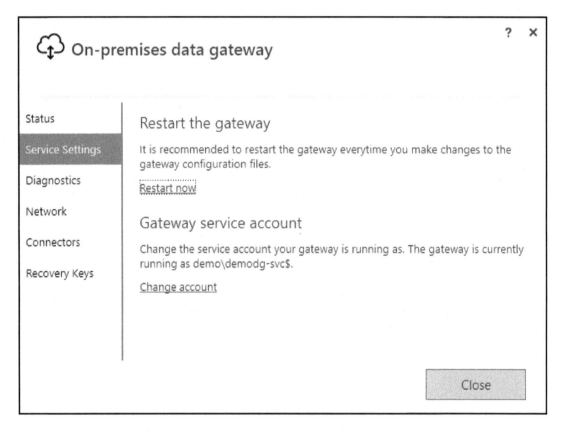

2. Under **Gateway service account**, select **Change account**.
3. In the dialog box, click **Change account** to confirm that you want to change the account.

4. Enter the name of the new account in the `DOMAIN\username` form. Enter the password and click **Configure**:

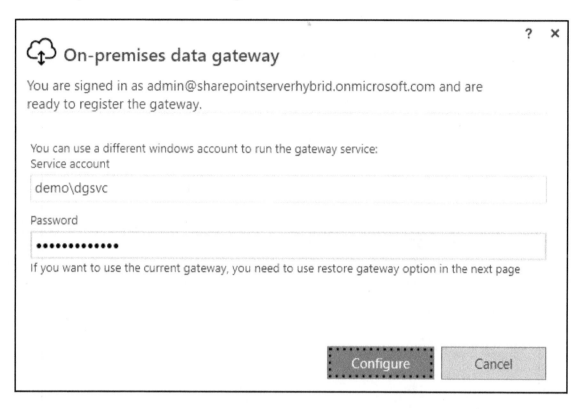

5. Wait as the gateway is updated.
6. When prompted, sign in as an administrator of this data gateway.
7. Select the **Migrate, restore, or takeover an existing gateway** option and click **Next.**

8. Enter the recovery key, select the correct instance (if you have a gateway cluster), and click **Configure**:

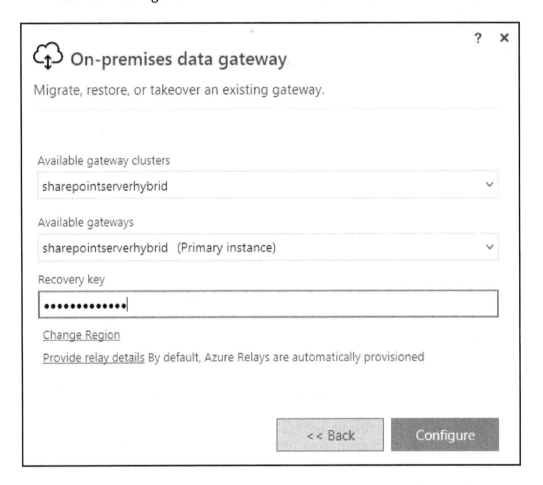

9. Wait as the gateway is recovered.
10. Verify that it has completed successfully and click **Close**:

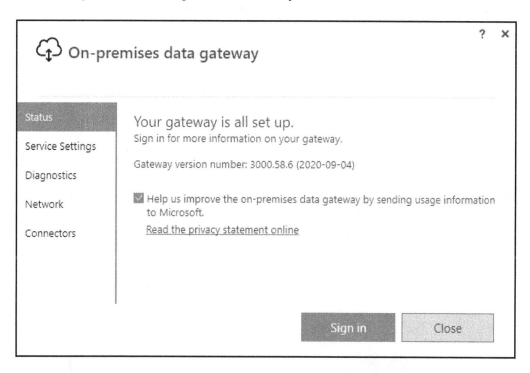

11. If running a gateway cluster, repeat the process on the additional nodes, selecting the corresponding gateway node/instance to update.

The update process is complete.

Changing the recovery key

The recovery key can be changed from the **On-premises data gateway** app configuration. To change the key, follow these steps:

1. From the computer that has the data gateway installed, launch the **On-premises data gateway** application.

2. Select the **Recovery Keys** tab and then select **Set new recovery key**:

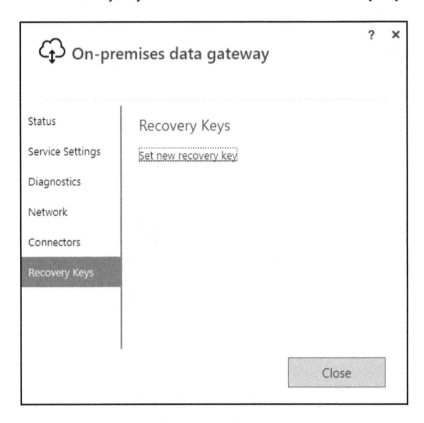

3. Select a data gateway. If you are updating the recovery key for a cluster, you need to select the primary instance. Enter the old key and the new key, and then click **Configure**:

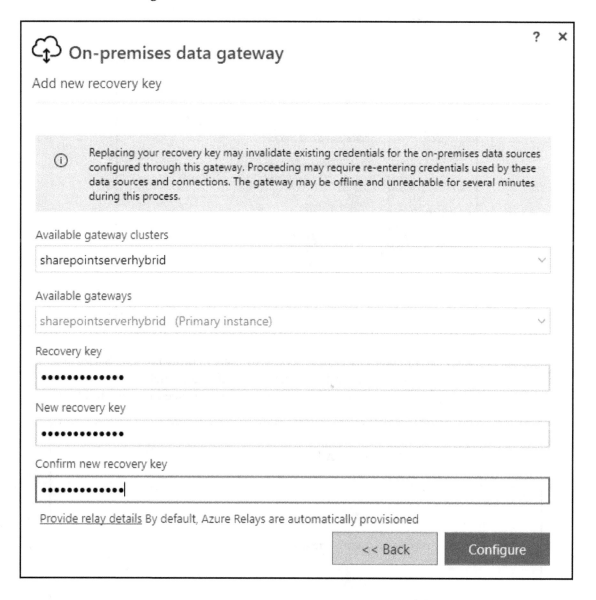

4. After it has been updated, you'll be returned to the main **On-premises data gateway** screen. Select **Recovery Keys** and notice that there are now options to **Delete legacy recovery key** (the old previous one) and **Set new recovery key**:

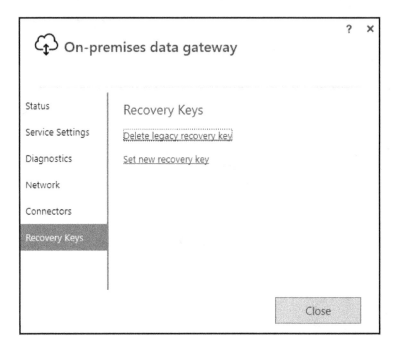

5. If you have a data gateway cluster, you will need to uninstall and perform a recovery on each of the gateway cluster nodes. If you launch the app without uninstalling first, you will receive a message that you need to uninstall first:

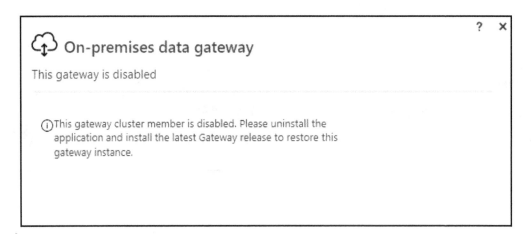

6. Uninstall, reboot, and re-install additional gateway cluster nodes.
7. After signing in, select **Migrate, restore, or takeover**.
8. Select the gateway cluster node you are recovering. Enter either the current recovery key or the previous (legacy) recovery key. Use only one of the keys:

Once you have finished changing the recovery key on the data gateways, you will need to update the data sources in the portal. To complete the update, follow these steps:

1. Log into the Power Platform admin portal (`https://app.powerbi.com/admin-portal`).
2. Select the gear icon from the menu bar and click **Manage gateways**.
3. Expand the cluster that was updated.
4. Select each data connection. The connection may display an error after updating nodes, as shown in the following screenshot:

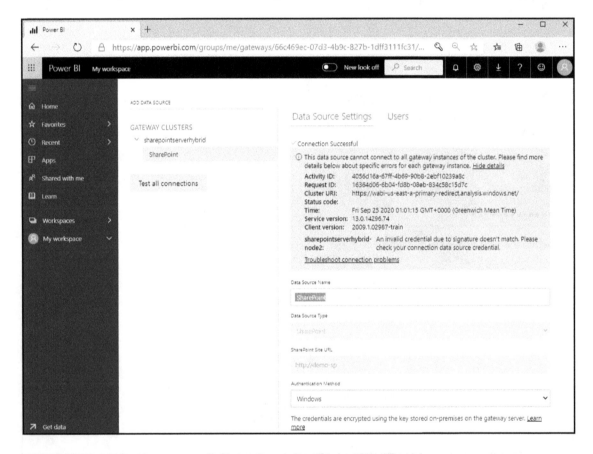

5. Re-enter the connection's credentials and save the gateway.

Repeat the preceding process for each data source and connection that the data gateway hosts.

Monitoring a gateway

Data gateways provide performance logging capability. You can use these features to help troubleshoot issues and narrow down the source of problems, such as inefficient queries or poor bandwidth.

Enabling gateway performance monitoring

Gateway performance can be monitored by enabling performance-logging capabilities in a configuration file. To enable the performance logging features, update `C:\Program Files\On-premises data gateway\Microsoft.PowerBI.DataMovement.Pipeline.GatewayCore.dll.config` with the following values:

1. Configure the `QueryExecutionReportOn` setting to `True` to enable additional logging for queries executed using the gateway. This option creates the `Query Execution Report` and `Query Execution Aggregation Report` files:

```
<setting name="QueryExecutionReportOn" serializeAs="String">
  <value>True</value>
</setting>
```

2. Configure the `SystemCounterReportOn` setting to `True` to enable additional logging for memory and CPU system counters. This option creates the `System Counter Aggregation Report` file:

```
<setting name="SystemCounterReportOn" serializeAs="String">
  <value>True</value>
</setting>
```

3. Update the values in the following table accordingly:

Setting	Description	Notes
ReportFilePath	Determines the path where the three log files are stored	The default configuration stores the report data in the user profile app data directory for the PBIEgwService service account. If you change this service account, you should update the path accordingly.
ReportFileCount	Determines the number of log files of each kind to retain	The default configuration is to retain 10 log files.
ReportFileSizeInBytes	Determines the size of the file to maintain	The default value is 104,857,600 bytes or approximately 100 MB.
QueryExecutionAggregationTimeInMinutes	Determines the number of minutes for which the query execution information is aggregated	The default value is 5 minutes.
SystemCounterAggregationTimeInMinutes	Determines the number of minutes for which the system counter is aggregated	The default value is 5 minutes.

4. After any changes to the configuration file are made, restart the gateway service.

You should begin seeing files in the location specified in the ReportFilePath value.

Enabling query logging

You can also troubleshoot slow-performing queries by enabling additional logging in the data gateway app.

To enable query logging, take the following steps:

1. On the computer hosting the data gateway, launch the **On-premises data gateway** app and select **Diagnostics**.
2. Enable the slider for **Additional logging** and click **Apply**:

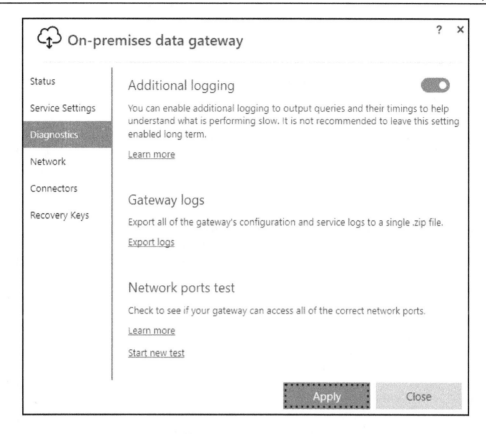

3. When prompted, click on **Apply and restart** to apply the settings and restart the gateway.

For more information on monitoring and visualizing a data gateway's performance using a Power BI template, refer to `https://docs.microsoft.com/en-us/data-integration/gateway/service-gateway-performance`.

Next, we'll look at how to troubleshoot some common issues.

Troubleshooting common issues

From time to time, it may become necessary to troubleshoot errors that occur in the data gateway. You can use the following procedures to resolve installation or functionality errors.

Communication errors

If changes are made to the network, you may experience communication issues. You can use the **Network ports test** option under the **Diagnostics** tab of the **On-premises data gateway** app to determine whether the data gateway can reach all of the necessary network endpoints:

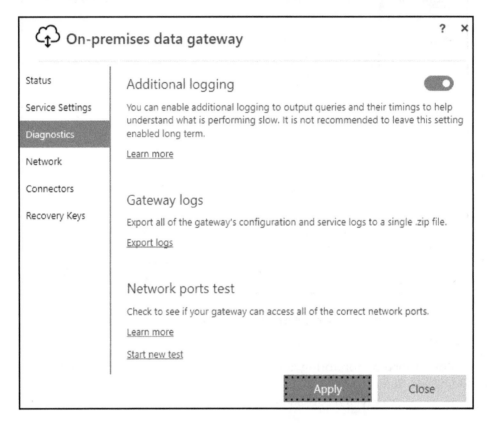

You can select **Start new test** to test the network communication to the internet from the data gateway.

If your network requires a proxy server to communicate with the internet, you may need to request a bypass for the service or configure the data gateway to use a specific user identity. You can use the steps under **Configure Proxy Server Authentication** to perform the necessary configuration changes.

Installation errors

If you are unable to install or update the data gateway app, you may need to perform installation troubleshooting. You may encounter a **Failed to add user to group. (-2147463168 PBIEgwService Performance Log Users)** error.

This error typically occurs if you are attempting to install the gateway on an AD domain controller. The data gateway service cannot be installed on a domain controller.

You may encounter a **File in use** error if you have on-access antivirus scanning configured on the computer where you are attempting to install the data gateway. Disable the antivirus software during the installation.

A **You are trying to reinstall a version already installed on the machine** error occurs if you are installing a version of the gateway service that is already installed. If you need to reinstall the current version, uninstall the existing gateway first:

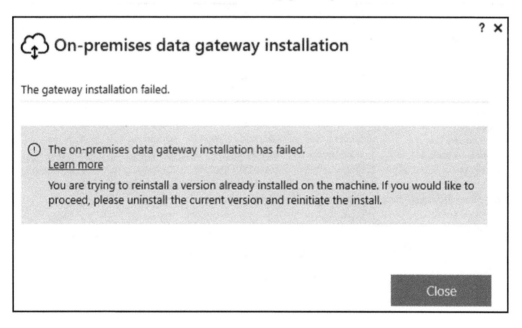

If you encounter any different errors to these, you may need to open a support ticket with Microsoft to get their assistance in its resolution.

Log sources

There are two main places to review log files for the data gateway.

First, you can use the **Diagnostics** tab of the data gateway app to export diagnostic logs. The logs are saved to the desktop of the logged-in user in a folder called ODGLogs. This logs folder contains individual log files for software installation setup components, network configuration, the computer environment, and the initial configuration.

You can also review the log item data in the event viewer. To view events logs for the gateway, follow these steps:

1. On the computer with the data gateway installation, launch **Event Viewer**.
2. Expand **Event Viewer | Applications and Services Logs** and select **On-premises data gateway service**:

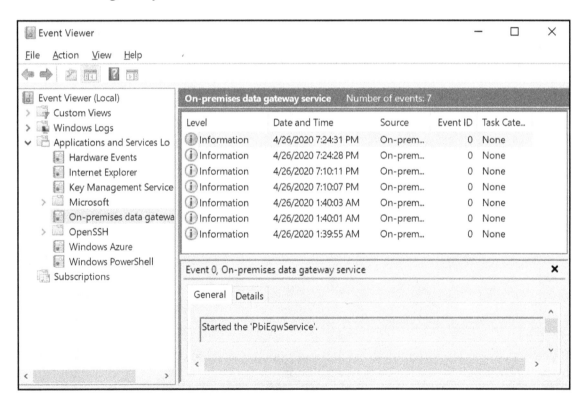

Reviewing the data in the previously shown log sources should help you resolve any issues. If you are unable to resolve an issue, contact Microsoft support.

Summary

After reading this chapter, you should be able to install and configure a data gateway, as well as perform common administrative tasks such as adding a data source, adding administrators, and changing the service account. We covered the networking requirements for the data gateway and the tools necessary to troubleshoot common installation problems. You should be able to successfully administer the data gateway and allow Office 365 services such as Power BI and Power Automate to connect to your on-premises data sources.

In the next chapter, we will introduce some basic concepts and planning for migrating on-premises SharePoint data to SharePoint Online.

Using Power Automate with a
Data Gateway

15

Power Automate is a new Microsoft platform for developing and managing workflows. With SharePoint Server 2019 (and previous versions), you can use SharePoint Designer and Workflow Manager to create business workflows to perform tasks such as automating document approvals. However, these tools are largely limited to working with data inside the SharePoint environment, reducing the scope of impact that a workflow can have.

With the Power Automate platform, workflows can be integrated with SharePoint and hundreds of other software-as-a-service-based applications, including popular products outside the Microsoft ecosystem. In fact, any service that has a REST-based interface can be used with Power Automate!

Because of this flexibility and power, Microsoft has begun emphasizing Power Automate as the preferred solution for creating and managing workflows.

The data gateway, introduced in `Chapter 14`, *Implementing a Data Gateway*, is the bridge to allow the Power Automate service in Microsoft 365 to access data resources located on-premises.

In this chapter, we're going to focus on the following topics:

- Learning about Power Automate
- Configuring Power Automate to use the data gateway
- Connecting Power Automate to SharePoint Server

By the end of this chapter, you should be able to describe Power Automate and configure Power Automate to use a data gateway.

Let's begin!

Learning about Power Automate

As mentioned previously, Power Automate is a workflow engine designed to give users the capability of automating business processes. Power Automate is a *flow* that comprises the following components:

- **Connectors** are components that are used to directly interface with both source and target applications or systems. Connectors contain the information required to interact with applications (such as the endpoint location and configuration or authentication information).
- **Triggers** are the activities that can initiate a flow. Triggers generally fall into three categories:
    - **Automated**: A webhook or notification method tells the flow to start based on a new event occurrence (such as a file being uploaded or a new database record being created).
    - **Instant**: A user initiates the flow by clicking a button.
    - **Scheduled**: The flow executes at defined time intervals.
- **Actions** describe the types of activities that are performed by a flow, such as copying a file, updating a record, or sending an email.

Power Automate is in a class of *citizen developer* tools, meaning that it's designed with end users in mind. Since it is user-oriented, you likely won't have a lot of administration interaction aside from working with the data gateway to allow information from SharePoint (or other on-premises systems) to communicate with Microsoft 365 and the Power Platform.

 You can learn more about working with Power Automate in the book *Workflow Automation with Microsoft Power Automate*, also available from Packt Publishing (`https://www.packtpub.com/product/workflow-automation-with-microsoft-power-automate/9781839213793`).

Next, we'll configure a connection inside Microsoft Power Automate to use the data gateway.

Configuring Power Automate to use the data gateway

Before you use Power Automate to access on-premises data, you'll need to configure a connection object inside the Power Automate interface. To configure connectivity, follow these steps:

1. Navigate to the Power Automate web portal (`https://flow.microsoft.com`) and sign in with an identity that has a Power Automate license assigned to it:

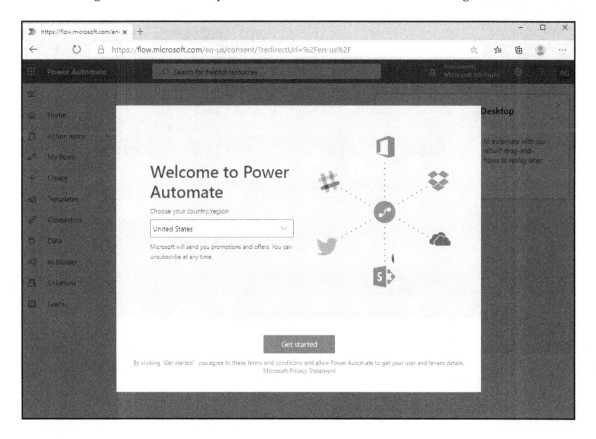

2. In the navigation bar, expand **Data** and select **Connections**:

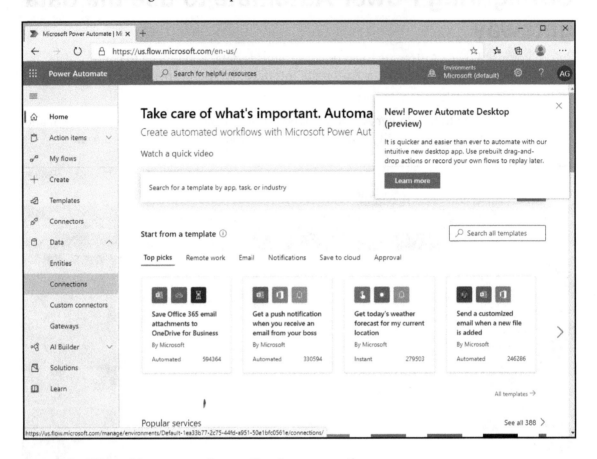

3. Click **+ New connection** or **Create a connection**:

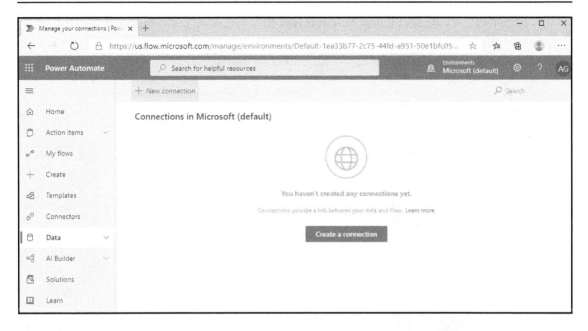

4. Select the type of resource you're connecting to that is located behind the gateway. In this instance, we're going to be working with SharePoint Server, so you'll want to select the **SharePoint** connection type:

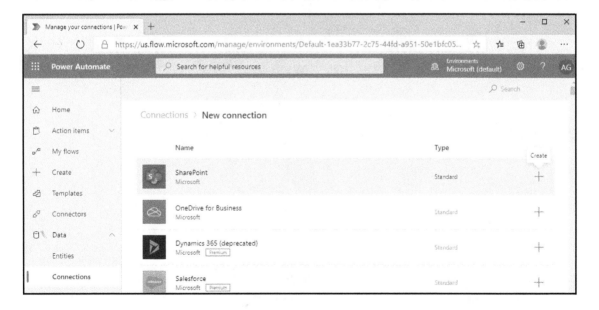

5. Select the **Connect using on-premises data gateway** radio button:

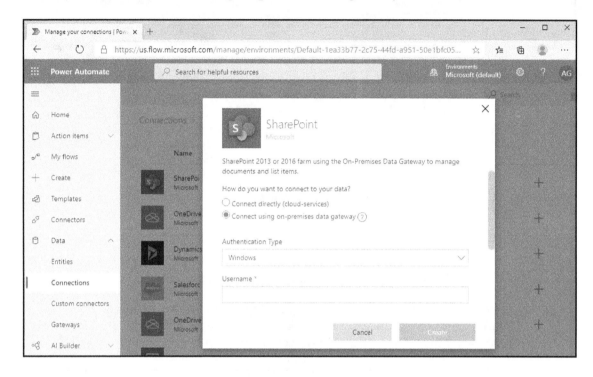

6. Enter a username (in the `DOMAIN\username` format) and a password. Select the appropriate gateway (if you have more than one). Click **Create**:

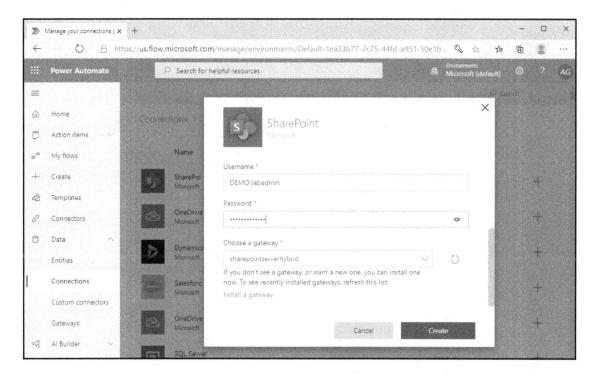

7. If the data gateway you have deployed has proper connectivity (as outlined in Chapter 14, *Implementing a Data Gateway*), the connection will be configured. Errors (such as improper credentials) will be displayed on this screen. Wait while the connection is added:

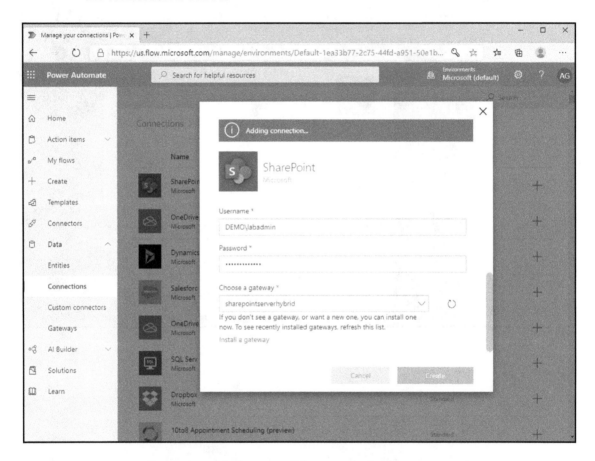

Once the connection has been added, you will be returned to the **Connections** page. You can then begin creating flows that utilize the gateway.

You can further explore the Power Automate data gateway configuration and options. Select **Gateways** under **Data** to see the data gateways available to use with Power Automate, as shown in the following screenshot:

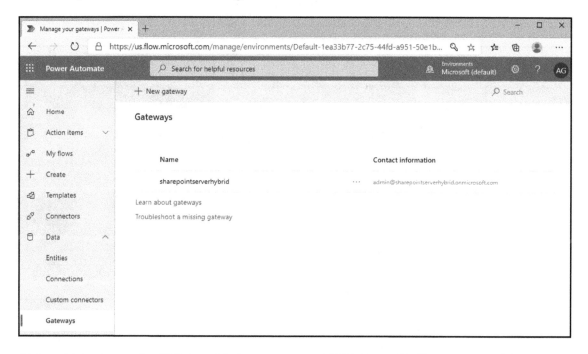

You can use this screen to view additional details about the gateway (such as the nodes that are participating in the gateway) and the user that originally configured the gateway.

In the next section, we'll use data entered in an on-premises SharePoint list as part of a flow.

Connecting Power Automate to SharePoint Server

Now that Power Automate is connected (via the data gateway) to the on-premises SharePoint Server, its environment can be used in any part of a Power Automate workflow.

In this example, we'll use Power Automate to send an email when a new item gets added to an on-premises SharePoint list. To work with this example, you'll need a SharePoint list (you can also use a document library, but will have to update the flow accordingly). You can use these steps to configure a simple flow to work with on-premises SharePoint Server data:

1. Navigate to the Power Automate web portal (`https://flow.microsoft.com`) and select **+ Create**:

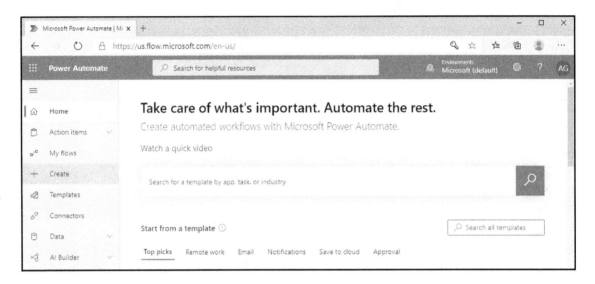

2. Select **Automated flow**:

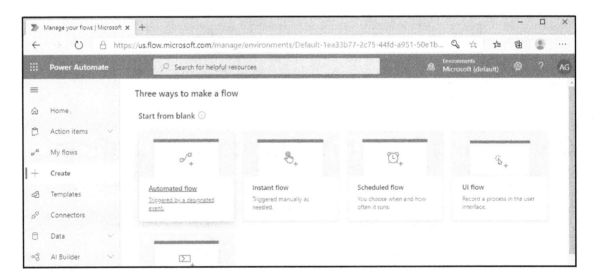

3. Add a descriptive name for the flow, and then select the **When an item is created** SharePoint trigger. Click **Create**:

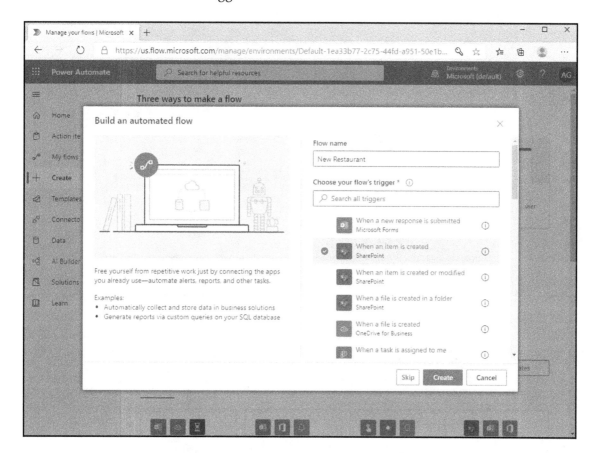

4. In the **Site Address** field, select the name of the site that contains the list. If the list does not populate (which is likely in this scenario), you'll need to enter the URL of the site that contains the list by selecting **Enter custom value**:

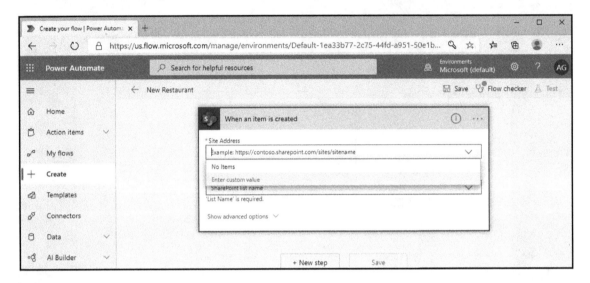

5. Select the list name:

6. Click **+ New step**.
7. In the **Choose an action** box, select the **Send an email (V2)** Outlook action:

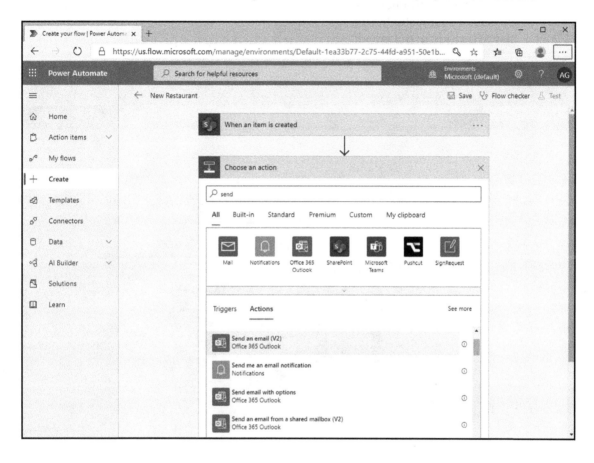

8. In the **To** field, enter the email address of a recipient. Enter a subject and body to complete the form. You can also use dynamic content tokens, which are values (or variables) automatically retrieved from the SharePoint site list that you selected previously. In this case, we selected the **Name** column of the SharePoint list to include in the email:

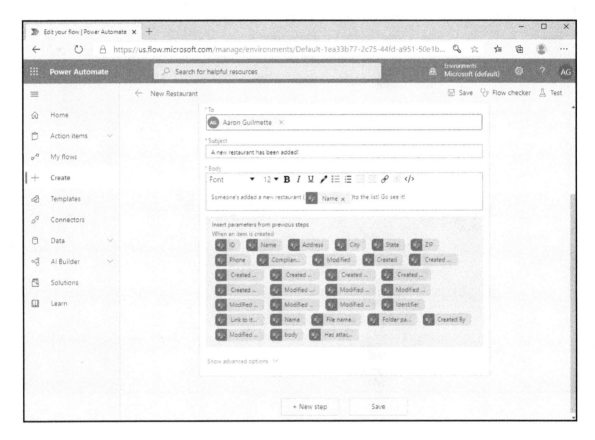

9. Click **Save**.

To test the flow, simply add a new item to the list, and then check the inbox of the recipient specified in the **To** box.

Summary

In this chapter, you learned how to use the Power Platform data gateway to provide access to on-premises data resources, such as SharePoint lists. You saw how to use these on-premises list components in a cloud-based workflow automation sequence. Connecting an on-premises environment via a data gateway is one of the best ways to expand your organization's automation capability.

In the next chapter, we'll begin reviewing data migration options and scenarios to help transition to SharePoint Online.

16

Overview of the Migration Process

Migrating content to SharePoint Online, whether you do so from local file shares, third-party file hosting services, SharePoint Server infrastructure, or even from another SharePoint Online tenant, requires a thoughtful approach. To successfully complete a SharePoint Online migration, you must undergo several steps, including the following:

- Determining and inventorying data sources
- Evaluating the migration process and tools
- Gathering data permissions and assessing the requirements
- Planning the destination site architecture and security
- Evaluating the network requirements
- Preparing data sources
- Change management

Each of these steps is important in ensuring data or applications aren't overlooked and that you have allocated adequate resources to the effort. Communication throughout the change management process is critical to driving the adoption of newly deployed technology and processes.

This chapter will introduce you to the planning concepts necessary to execute a successful migration.

Let's go!

Determining and Inventorying Data Sources

After choosing to migrate to SharePoint Online, one of the first steps in building a migration strategy is to identify what data needs to be migrated. This can include the following:

- Group- or team-networked file shares on file servers or network-attached storage
- SharePoint Server document libraries and lists
- SharePoint Server applications
- SharePoint Server workflows
- InfoPath forms
- Intranet content

For each of these content types, you need to evaluate whether this content is fit to migrate to SharePoint Online or whether another product, service, or migration path may better support the business model. The following diagram shows potential data sources and targets:

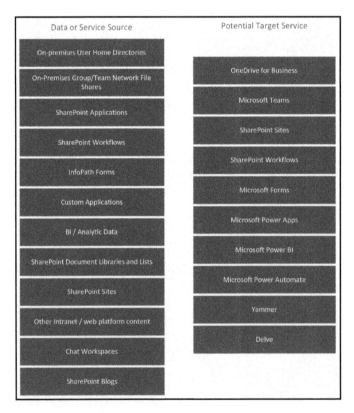

Considerations must be made for how users currently interact with data and components, as well as business process and application support scenarios.

When planning a SharePoint Online migration process, you'll need to map out which target online services best support the business processes. You'll need to plan for training and accommodating line-of-business applications. Finally, you'll need to inventory existing applications to plan for any additional costs associated with upgrading or reconfiguring them.

Once you have an understanding of what needs to migrate and what the future service mapping looks like, you can begin evaluating the methods and procedures for executing the migration.

Evaluating the Migration Process and Tools

There are many methods available for migrating data to SharePoint Online and Office 365, including both self-service and assisted options. Depending on an organization's budget and technical capability, the amount of data, source environments, and the timeline, they may decide to use one or more of these methods.

Self-service

With self-service tools, organizations can complete migrations to Office 365 and SharePoint Online themselves. The Microsoft services listed here are mostly free of charge (the exception being the freight, shipping, and hardware fees for Azure Data Box). Many organizations use these self-service methods with great success, but they do require a certain amount of technical skill, planning, and work effort:

- **The SharePoint Migration Tool** (**SPMT**): This is a Microsoft tool used to migrate content from file shares or existing on-premises SharePoint document libraries and lists. You can learn more about SPMT at `https://spmtreleasescus.blob.core.windows.net/install/default.htm`.
- **SharePoint Migration Manager**: This tool is a user interface-driven self-service tool from within the SharePoint Online admin center. It uses agents that are installed on-premises to migrate data. SharePoint Migration Manager is currently in preview. You can learn more about SharePoint Migration Manager at `https://docs.microsoft.com/en-us/sharepointmigration/mm-get-started`.

- **Azure Data Box**: The Azure Data Box solution is suitable for organizations whose content to migrate to SharePoint Online is measured in terabytes as opposed to gigabytes. Azure Data Box is a combination of a device and a service: Microsoft ships a device to you, you transfer the data to it, and then you ship it back to Microsoft, where it is installed on a data center and then ingested into your tenant. As this method involves hardware, there is a fee for this service. For more information on Azure Data Box solutions, see `https://docs.microsoft.com/en-us/sharepointmigration/how-to-migrate-file-share-content-to-spo-using-azuredatabox`.

- **Mover**: The newest addition to this list, Mover is a third-party application that Microsoft acquired in October 2019 to assist customers in workload and data migration. While it *can* be used to move on-premises file shares to Office 365, it is currently not supported when migrating from on-premises SharePoint to SharePoint Online. For more information on Mover, see `https://docs.microsoft.com/en-us/sharepointmigration/mover-fileshare-to-o365`.

- **Third-party tools**: There are also a number of third-party tools on the market, including offerings (in no particular order) from AvePoint, ShareGate, Metalogix, BitTitan, and many others. Some third-party tools offer included (or paid) support options, support for additional content types or sources, scheduling, and additional reporting capabilities.

While this book doesn't focus on third-party tools for the purposes of migration, it's important to know that they exist and can provide some additional features that the native tools may not.

Organizations with more complex needs, however, frequently choose to use service offerings.

Service engagements

In addition to the previously mentioned self-service options, there are also service engagements provided by Microsoft and their partners to assist customers in migrating to Office 365 and SharePoint Online:

- **Microsoft FastTrack**: Microsoft FastTrack is a service offering provided free of charge by Microsoft to customers with at least 150 user licenses. FastTrack is a remote assistance offering and typically provides guidance, with the customer performing many of the tasks. You can learn more about the FastTrack services at `https://docs.microsoft.com/en-us/fasttrack/o365-data-migration`.

- **Microsoft Consulting Services**: A Microsoft Consulting Services engagement is a full-service, custom endeavor involving a team of service professionals from Microsoft. A Services team typically includes delivery and project managers, architects, and consultants. You can learn more about the Microsoft Consulting Services portfolio at `https://www.microsoft.com/en-us/industry/services/consulting`.

- **Microsoft Partner services**: Microsoft Partner can offer a full range of consulting engagements, including custom development and integration with third-party applications. Through Microsoft's worldwide partner network, organizations can find industry-specific consultancies to help guide them on their Office 365 journey. You can learn more about the Microsoft Partner network or find partners by visiting `https://www.microsoft.com/en-us/solution-providers/home`.

Whether you choose to perform migration activities yourself or engage with outside resources for some (or all) of the process, you'll want to be aware of your options. Many service offerings will also use third-party tools, and those may generate additional costs that should be considered when planning and budgeting a migration project.

Regardless of which options you choose, you'll likely need to be involved in evaluating the source and destination environments.

Gathering Data Permissions and Assessing the Requirements

Many organizations use Windows network file shares or **Network-Attached Storage** (**NAS**) solutions that rely on a combination of share and NTFS permissions to manage access. SharePoint sites, lists, and libraries similarly use a combination of explicit and inherited permissions (at the collection, site, library, and item level) to control access to resources. Other applications and services may use explicit app-defined permissions, rely on Azure **Active Directory** (**AD**) or local group permissions, or may use another access control mechanism altogether.

When moving to SharePoint Online, it's important to understand how these currently work so that you can plan for how permissions will work in the future. Many tools provide some sort of mapping between on-premises security controls and SharePoint Online security controls. This can be automated through the tool's interface or through the use of a separate mapping file.

For example, SPMT allows you to use native Azure AD synchronization to populate a mapping table or for the administrator to provide a mapping file.

As an example, the following table lists how SPMT handles permissions when migrating content to a SharePoint site, libraries, or lists:

Source	The Preserve user permissions Setting	Migration Destination	Target Permission before Migration	Target Permission after Migration	Notes
File share	Off	Root folder	Inherited	Inherited	The role assignments of the target library's existing files won't be changed; migrated files have **Inherited** permissions (that is, they have inherited role assignments from the target library).
File share	Off	Root folder	Unique	Unique	
File share	Off	Subfolder	Inherited	Inherited	
File share	Off	Subfolder	Unique	Unique	
File share	On	Root folder	Inherited	Unique	The role assignments of the target library will be replaced by those in the source root folder. Existing files with inherited permissions will still be inherited permissions but with a new role assignment from the target library. Existing files with **Unique** permissions won't be changed. Migrated files without explicit permissions in the source will have inherited permissions and inherited role assignments from the target library. Migrated files with any permissions in the source will carry over these permissions as unique.

Source	The Preserve user permissions Setting	Migration Destination	Target Permission before Migration	Target Permission after Migration	Notes
File share	On	Root folder	Unique	Unique	Permissions from the source folder will be added as new role assignments to the target library. Existing files with inherited permissions will still be inherited permissions but with a new role assignment from the target library. Existing files with unique permissions won't be changed. Migrated files without any permissions in the source will have inherited permissions and inherited role assignments from the target library. Migrated files with any permissions in the source will carry over these permissions as unique.
File share	On	Subfolder	Inherited	Inherited	Role assignments of the target library and existing files won't be changed. Permissions from the source folder and files will be carried over to the target subfolder and corresponding files, which will have **Unique** permissions as new role assignments.

Source	The Preserve user permissions Setting	Migration Destination	Target Permission before Migration	Target Permission after Migration	Notes
File share	On	Subfolder	Unique	Unique	Role assignments of the target library and existing files won't be changed. Permissions from the source folder and files will be carried over to the target subfolder and corresponding files that will have **Unique** permissions as new role assignments.
List/document library	Off	Root folder	Inherited	Inherited	
List/document library	Off	Root folder	Unique	Unique	
Document library	Off	Subfolder	Inherited	Inherited	
Document library	Off	Subfolder	Unique	Unique	Treated the same as file share migration with the corresponding conditions.
List/document library	On	Root folder	Inherited	Unique	
List/document library	On	Root folder	Unique	Unique	
Document library	On	Subfolder	Inherited	Inherited	
Document library	On	Subfolder	Unique	Unique	
Site/web	Off	NA	Inherited	Inherited	The role assignment of the target site/web will be unchanged.
Site/web	Off	NA	Unique	Unique	
Site/web	On	NA	Inherited	Unique	The role assignment of the target site/web *will be replaced* by those in the source site/web.

Source	The Preserve user permissions Setting	Migration Destination	Target Permission before Migration	Target Permission after Migration	Notes
Site/web (A) with subsite B (both migrated with SPMT)	On	NA			Site A will follow normal site migration with the same settings. Subsite B will become unique and the role assignment *will be replaced* by those in the source of subsite B.
Site/web	On	NA	Unique	Unique	The role assignment of the source site/web will be added as new role assignments to the target site/web.

As you can see from the preceding table, inventorying and gathering source permissions and working out *who* should have access to *what* can be a complex task.

Many organizations choose not to preserve existing file share or site permissions when migrating to SharePoint Online, which can have both pros and cons. Attempting to preserve source permissions can help you manage expectations for access control when migrating, but it could potentially come at a significant cost during the discovery and planning stages when trying to evaluate permissions. Each organization must ultimately decide how and where to invest its time and what the potential return on investment or security is for preserving file and site permissions.

In the next section, we'll look at planning the **destination** site architecture and security—an activity that, if done well, can reduce the amount of time necessary to invest in permissions discovery.

Planning the Destination Site Architecture and Security

If we look at the previous table, it's easy to see how complex site and file permissions can get. Every organization must address permissions and architecture in the *new* environment, and a migration event is a good opportunity to "start fresh."

SharePoint Online allows organizations to continue using the classic site architecture components (such as site collections and subsites), but also provides a new architecture paradigm called **hub site architecture** to help simplify design choices and provide more flexibility. If you decide to shift to modern site architecture, you'll need to ensure that the new destination sites have the required explicit memberships.

Hub site architecture is based on modern SharePoint sites, which maintain security and access control through an Office 365 group membership connected to the site. You can learn more about developing modern sites at `https://docs.microsoft.com/en-us/sharepoint/information-architecture-modern-experience`.

Next, we'll discuss how to start planning network requirements.

Evaluating Network Requirements

Migration to an online service such as SharePoint Online requires two aspects of network considerations:

- The network resources required to *migrate* content and applications
- The network resources required for ongoing operational activity

We'll look at the considerations for both.

Planning the migration requirements

When determining the network requirements for a migration event, you'll need to have accurate information on how *much* data you're migrating as well as the *speed* at which you can migrate.

The volume question is relatively easy to surmise—you'll be tallying up file shares, home directories, intranet sites, existing SharePoint site collection data, and any other application data that will be in scope for a migration effort. Network device planning should cover things such as the following:

- **Proxy devices**: Microsoft recommends that you bypass proxy devices or applications for traffic going to Office 365. If your network environment utilizes proxy infrastructure, you'll need to make sure that the migration traffic is excluded from it (per the Microsoft recommendations at `https://docs.microsoft.com/en-us/microsoft-365/enterprise/networking-configure-proxies-firewalls`) or that you scale your infrastructure accordingly to be able to handle the additional load.
- **Stateful packet inspection firewalls**: Since all traffic is SSL-encrypted, inspection efforts will require the decryption and re-encryption of packets. Large migration efforts will likely put a strain on these resources as they perform other normal daily operations.
- **Intrusion protection or intrusion detection devices**: In both theory and practice, the Microsoft 365 data center environment is an extension of your environment. You may want to continue to run the **Intrusion Detection System** (**IDS**) monitoring, but at least for purposes of migration, disable any **Intrusion Prevention System** (**IPS**) from taking any action. Migration to cloud services is technically a data exfiltration event. IPS that intervenes in-network access can interfere with a smooth migration.

Migration bandwidth and resources can be thought of as a type of peak load—you likely won't tax your infrastructure as much as you will during this type of event. However, once you've migrated, you need to plan for the reality of how your network infrastructure will be impacted during the normal daily business cycle.

That's where network requirements related to operations come into play.

Planning the operational requirements

When looking at requirements planning through an operational lens, you'll be trying to identify how your network will be utilized during the regular business activity—including the spikes or peaks that happen, such as month-end or year-end processing.

One of the best places to gather this information is by using network analysis of your current file-serving environments, specifically looking at daily and peak trends for traffic between user/client LAN segments and the servers hosting group, team, and home directory data. You'll want to exclude traffic to/from any backup environment as that will skew the data transfer volume estimates.

Additionally, Microsoft has provided a new proof-of-concept tool, the **Office 365 Network Onboarding Tool**, to help gather information about your users. This tool can be used for both migration and the operational requirements for data gathering. The tool is located at `https://connectivity.office.com/`:

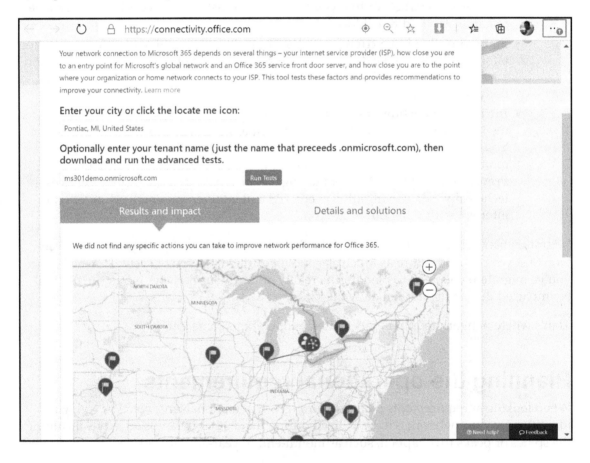

This tool can provide suggestions as to how to improve both your migration and operational experiences. You may also find it useful to use tools such as NetMon or Wireshark to help capture and evaluate network traffic.

After gathering information on the data you intend to migrate and for your migration environment, you'll need to look at the steps you can take to prepare your data for migration.

Preparing Data Sources

Preparing data for migration can involve several steps, each of which has unique caveats depending on your environment:

- Review and update any filenames and pathnames that are not compliant with SharePoint Online.
- Determine whether any data isn't a candidate for SharePoint Online (such as prohibited file types or extensions).
- Review metadata and taxonomy configurations.
- Back up any customizations that may need to be manually deployed.
- Back up any site templates and site customization scripts.
- Review currently deployed third-party apps or plugins that need to be upgraded or replaced.
- Determine an action plan for any plugins or apps that haven't got a replacement or upgrade path.

The steps to prepare source data will vary based on your environment. Most organizations will have to go through exercises to identify content that needs to be modified or updated in some way to transition smoothly to SharePoint Online.

SPMT, as well as many third-party tools, has scanning components to identify many of these issues. Additionally, SPMT (and third-party tools) can migrate files, sites, pages, version information, permissions, and a large number of customizations.

For an up-to-date list of the features that SPMT can migrate out of the box, see `https://docs.microsoft.com/en-us/sharepointmigration/what-is-supported-spmt`.

The final step in preparing for migration is change management, which we will cover next.

Change Management

It's been said before that in technology, the only constant is change. Many organizations struggle with concepts of change management, especially when dealing with such highly integrated pieces of technology, such as SharePoint and document management platforms.

Whether embarking on a SharePoint Online migration internally, using partner resources, or engaging with the Microsoft FastTrack center, developing an adoption plan is key to getting value out of your migration effort.

Microsoft's adoption and change management framework can be separated into three key areas, each of which has multiple components:

- **Envision**: In the envisioning stage, you identify key stakeholders in the business and determine particular business scenarios to enable. For example, a key stakeholder may be the CTO's office and their business scenario might be enabling access to resources currently stored on-premises in the event of a disaster that requires people to work from home. Then, you can start building a plan that helps define what success looks like for that particular business scenario. A success plan typically has criteria and **Key Performance Indicators (KPIs)** that can be used to objectively determine whether the goal has been met.
- **Onboard**: In the onboarding phase, you are deploying or migrating both technology assets *and* people. Onboarding activities will typically involve an early adoption program to help you identify migration issues and fine-tune the materials that you'll use to communicate later on. You'll develop pre-launch, launch, and post-launch action plans to help keep the stakeholders, technologists, and early adopters in the loop as to the progress of activities and what they should be expecting.
- **Drive value**: Based on feedback, performance, and other analysis, you'll determine when your organization is ready to push further into the rollout. Using the insights from the early adopter program, along with training and communications material, you can show users the value of the platform and get them engaged in using it.

Any successful adoption and change management initiative focuses on showing the users how this change will benefit them. You can see samples of the change management, communications, and success plans on the Microsoft Change Management Framework page at `https://www.microsoft.com/microsoft-365/partners/changemanagementframework`.

Summary

In this chapter, we covered an overview of planning for a successful migration to SharePoint Online, such as determining bandwidth requirements and what type of permissions strategy you're going to employ. We reviewed, at a high level, SharePoint Online site architecture ideas (such as hub sites and modern sites) and how you might map your current design into that.

We also looked at some of the migration tools and paths you can choose, such as SPMT or utilizing Microsoft partner resources.

Finally, we addressed one of the most important topics—change management.

In the next chapter, we'll look at the steps to actually perform migrations using the native tools.

17
Migrating Data and Content

In this final chapter, we're going to look at processes, tools, and strategies for migrating on-premises SharePoint Server content to Office 365 and SharePoint Online.

SharePoint migrations require additional planning and tools not included with the product, some of which were covered in Chapter 16, *Overview of the Migration Process*. SharePoint migrations, as we discussed in the previous chapter, involve inventorying content to be migrated and mapping business priorities to the services that will be enabled.

In this chapter, we'll work through the following tasks and concepts:

- Prerequisites
- Scanning SharePoint content for migration issues
- Identifying and resolving blocking issues that prevent migration
- Determining the course of action for data that cannot be migrated
- Planning and configuring an automated SharePoint migration
- Recommended tools and strategies to migrate data

The most popular and full-featured tool that Microsoft provides is the **SharePoint Migration Tool** (**SPMT**). In the sections relating specifically to migration tooling and procedures, we'll use this tool as an example. While there are Microsoft-provided tools (such as SharePoint Migration Manager and Mover.io), they currently lack the ability to address as many migration scenarios as SPMT can.

By the end of this chapter, you'll be familiar with planning and conducting migrations to SharePoint Online.

Prerequisites

Before you start using SPMT to migrate content to SharePoint Online, you'll need to have followed the steps in Chapter 16, *Overview of the Migration Process,* to identify the data sources (such as SharePoint content or file shares) to migrate.

In the next sections, we'll review the system requirements (both on-premises and SharePoint Online), as well as network requirements for the successful communication and migration of data.

Network

In order for migrations to be successful, you will need to work with your organization's network team to ensure the server running the migration tool is able to connect to the following endpoints:

Site	Purpose
https://secure.aadcdn.microsoftonline-p.com	Authentication
https://login.microsoftonline.com	Authentication
https://api.office.com	Microsoft 365 **application programming interfaces** (**APIs**) to move and validate content
https://graph.windows.net	Microsoft 365 APIs to move and validate content
https://spmtreleasescus.blob.core.windows.net	Installation
https://*.queue.core.windows.net	Migration API (Azure requirement)
https://*.blob.core.windows.net	Migration API (Azure requirement)
https://*.pipe.aria.microsoft.com	Telemetry/update
https://*.sharepoint.com	Destination for migration
https://*.blob.core.usgovcloudapi.net	Migration API (Azure Government requirement)

`https://*.queue.core.usgovcloudapi.net`	Migration API (Azure Government requirement)
`https://spoprod-a.akamaihd.net`	**User interface (UI)** icons
`https://static2.sharepointonline.com`	UI icons

Now, let's dive into the prerequisites of Sharepoint Online in the next section.

SharePoint Online

Some web parts may require scripting in order to migrate successfully. You will need to ensure scripting is enabled in SharePoint Online, at least for the duration of the migration. To enable scripting, follow these steps:

1. Navigate to the Microsoft 365 admin center (`https://admin.microsoft.com`), expand **Admin centers**, and then select **SharePoint**.
2. Select **Settings**, and then click the link for the **classic settings page**.
3. Scroll down and select the **Allow users to run custom script on personal sites** and **Allow users to run custom script on self-service created sites** radio buttons, as illustrated in the following screenshot:

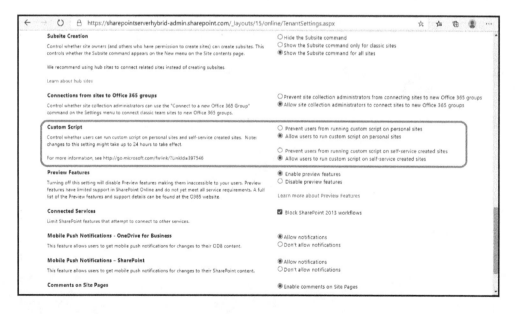

4. Click **OK** to save the settings.

These settings should be updated 24 hours before starting migration activities.

SharePoint Migration Assessment Tool

You also will want to obtain the **SharePoint Migration Assessment Tool** (**SMAT**) to help scan for potential issues. Use the following process to obtain the tool:

1. Navigate to `https://www.microsoft.com/download/details.aspx?id=53598`.
2. Download the ZIP file and extract it to a location on your computer.

SMAT also provides an identity assessment, which will help you map user identities for SharePoint Online.

SPMT

Finally, you'll need to download SPMT at `https://aka.ms/spmt-ga-page`. For ease of use, download and install the tool on a SharePoint server using the following steps:

1. Navigate to `https://aka.ms/spmt-ga-page`.
2. Click **Install** to download the tool, as illustrated in the following screenshot:

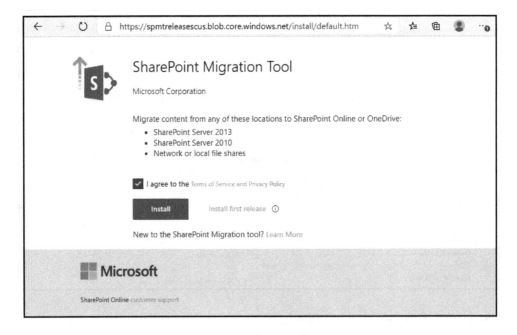

3. After the installer downloads, click **Open file** to launch the installer.
4. Provide Office 365 global administrator credentials to sign in to the tool prior to installation.
5. If you are not yet ready to conduct a migration, you can close the tool. You can relaunch it using the icon on the desktop, shown in the following screenshot:

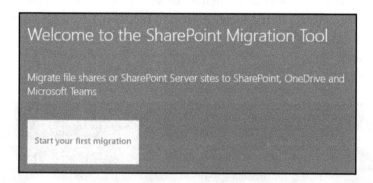

Welcome to the SharePoint Migration Tool

Migrate file shares or SharePoint Server sites to SharePoint, OneDrive and Microsoft Teams

Start your first migration

After SPMT is ready, you can begin scanning your existing content.

Scanning SharePoint content for migration issues

SMAT has two core functions, as follows:

- Generating an identity mapping file for existing sites
- Scanning content for potential issues

We'll look at each of these steps in the next sections.

Generating an identity mapping file for existing sites

To generate an identity mapping file, follow these steps:

1. From a SharePoint server, launch an elevated Command Prompt.
2. Change directories to the location from which SMAT was extracted.
3. Run the following command:

   `SMAT.exe -generateidentitymapping`. This is illustrated in the following screenshot:

```
C:\SMAT_RTM_Update6_1>smat.exe -generateidentitymapping

Generates identity mapping reports based on information from the current SharePoint farm, Active Directory,
and Azure Active Directory. It is required to provide consent for this application to read your Azure Active
Directory instance. If you do not have Azure Global Administrator rights to provide consent to the
application at runtime, an Azure Global Administrator will need to run SMAT.exe -ConfigureIdentityMapping
before a SharePoint admin is able to run SMAT.exe -GenerateIdentityMapping. For more information, see
http://aka.ms/SPIdentityMappingTool. The scan engine will start in a new console window and this program will
exit.

Launching: SMIT.exe -generateidentitymapping

Thank you for using the SharePoint Migration Assessment Tool!

C:\SMAT_RTM_Update6_1>_
```

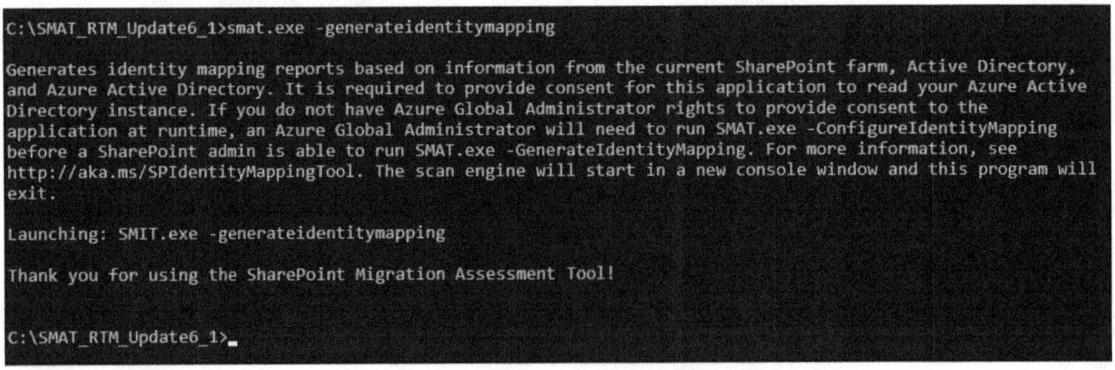

4. When prompted, enter an Office 365 global administrator credential.
5. Select the box to grant consent and click **Accept**, as illustrated in the following screenshot:

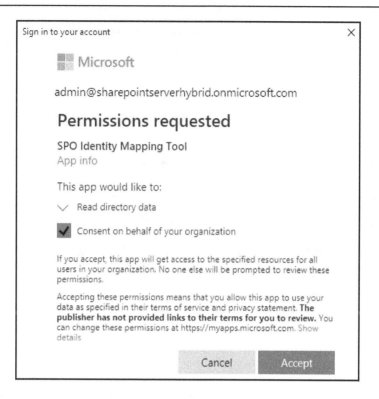

6. Review any errors generated by the identity mapping tool. The output from this is illustrated in the following screenshot:

7. Review the output file for any identities that are not synchronized to **Azure Active Directory** (**Azure AD**). To do so, open the output file (its default name is `FullIdentityReport.csv`) in the `Log` subfolder where the tool was run. Locate identities where the **TypeOfMatch** column indicates **NoMatch**, and review whether they are necessary for SharePoint Online. Repeat the process for items listed as **PartialMatch**. The output file is illustrated in the following screenshot:

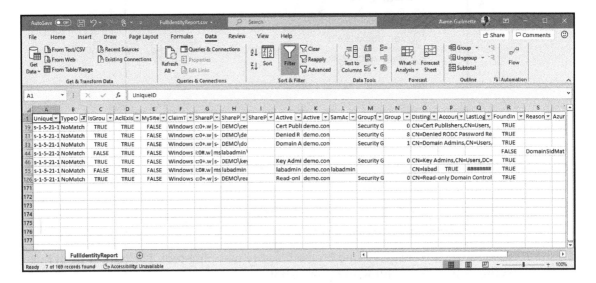

8. Make any updates in the AD or Azure AD to ensure the necessary files are matched. Repeat the identity mapping and review process until the identity report matches your expectations.

The most common issue you will encounter will likely be that no target identity exists in Azure AD for a corresponding on-premises account. This may be due to issues such as an Azure AD Connect filter or a duplicate on-premises identity constraint preventing synchronization. For a complete list of reason codes causing a **NoMatch** condition, see `https://docs.microsoft.com/en-us/sharepointmigration/sharepoint-migration-identity-mapping-tool`.

The `IdentityMapping.csv` file indicates users and groups that can be used to map an on-premises identity to the corresponding cloud identity.

Once identity mapping procedures are complete, you can move on to scanning for content issues.

Scanning content for potential issues

The next step in preparing for migration is to identify any content- or feature-based issues that could impact the process. You can use SMAT for this as well.

To launch SMAT and scan your environment, proceed as follows:

1. Log in to SharePoint Server as either the **Farm Service** account or as **Farm Administrator** (as long as the farm administrator has access to all web applications).
2. Launch an elevated prompt and change to the directory from which SMAT has been extracted.
3. Run the following command:

   SMAT.exe

   The process will identify the version of the tool to use (SharePoint Server 2010 or 2013, or later) and launch the appropriate scanner.

4. After the scanning has completed, you will be prompted to enter your SharePoint Online tenant, as illustrated in the following screenshot:

```
SharePoint Migration Assessment Tool version 1.0.1804.23001

Scan                              Status     % Complete  # Items     Status Message
===============================   =========  ==========  ==========  ======================================================
Alerts                            Finished   100         0           Finished scan work
Apps                              Finished   100         0           Finished scan work
BCSApplications                   Finished   100         5           Finished scan work
BrowserFileHandling               Finished   100         0           Finished scan work
CheckedOutFiles                   Finished   100         1           Finished scan work
CustomizedPages                   Finished   100         10          Finished scan work
CustomPermissionLevel             Finished   100         0           Finished scan work
CustomProfilePropertyMappings     Finished   100         1           Finished scan work
EmailEnabledLists                 Finished   100         0           Finished scan work
ExternalLists                     Finished   100         2           Finished scan work
FileVersions                      Finished   100         0           Finished scan work
FullTrustSolution_Farm            Finished   100         0           Finished scan work
FullTrustSolution_Content         Finished   100         0           Finished scan work
InfoPath                          Finished   100         0           Finished scan work
IRMEnabledLibrary                 Finished   100         0           Skip scanner work because IRM is not enabled in t...
LargeExcelFiles                   Finished   100         0           Finished scan work
LargeLists                        Finished   100         0           Finished scan work
LargeListViews                    Finished   100         0           Finished scan work
LargeSites                        Finished   100         0           Finished scan work
LockedSites                       Finished   100         0           Finished scan work
LongODBUrl                        Finished   100         0           Finished scan work
ManagedMetadataLists              Finished   100         106         Finished scan work
NonDefaultMasterPages             Finished   100         0           Finished scan work
PublishingPages                   Finished   100         0           Finished scan work
PublishingSites                   Finished   100         4           Finished scan work
SandboxSolution                   Finished   100         0           Finished scan work
SecureStoreApplications           Finished   100         0           Finished scan work
SiteTemplateLanguage              Finished   100         0           Finished scan work
UnsupportedWebTemplate            Finished   100         3           Finished scan work
WebApplicationPolicy              Finished   100         7           Finished scan work
WorkflowAssociations2010          Finished   100         0           Finished scan work
WorkflowAssociations2013          Finished   100         0           Finished scan work
WorkflowRunning2010               Finished   100         0           Finished scan work
WorkflowRunning2013               Finished   100         0           Finished scan work


Generating final report...       100%        17/17

Report data and log files can be found at:
C:\SMAT_RTM_Update6_1\Log


Optional: If you have already signed up for Office 365 and know the SharePoint Online tenant domain
url that you want to use for migration, you can include that URL to help us improve our services.

Would you like to provide your SharePoint Online tenant domain url?
Enter [Y]es or [N]o then press enter: _
```

After the reports are finished, you can review them in the Log subfolder.

SMAT can also be customized using the following configuration files:

- SiteSkipList.csv: This configuration file is used to list sites that you do not wish to include in the output.

- `ScanDef.json`: This configuration file is used to enable or disable individual scans for SMAT. You can edit the file, changing the `Enabled` parameter to `True` or `False` to enable or disable an individual scan. Some scans have additional options or properties available.

In addition, you can use SPMT to scan content for other issues that may develop during migration (such as invalid characters in filenames or file size issues). You'll see that option when we step through the migration process.

Next, we'll investigate issues that could impact or prevent a successful migration.

Identifying and resolving blocking issues that prevent migration

To identify items that need to be resolved prior to migration, you can review the log output from SMAT. There are three core logs located in the `Log` output directory, as follows:

- `SMAT.log`: General log file, containing information, warning, and error messages. Error conditions should be investigated, as they indicate issues that may prevent the user from moving forward.
- `SMAT_Errors.log`: Log file containing only errors. If the file is not present after running SMAT, then no errors were present.
- `SMATTelemetry.log`: Data for telemetry; does not affect the migration process or reporting.

An example of the `SMAT_Errors` log is shown in the following screenshot, referencing that an unauthorized access exception was returned:

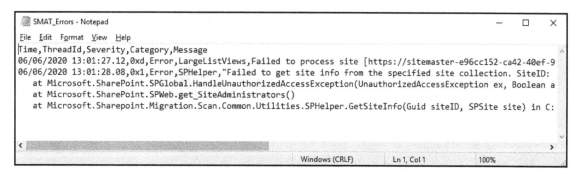

Using the error messages in the log file, resolve any system issues. You may want to continue rerunning SMAT, attempting to resolve errors until the error log is clear.

Detailed information about the results of scans is available in the `Log\ScannerOutput` folder. You can compare those outputs to the diagnostic details listed at `https://docs.microsoft.com/en-us/sharepointmigration/sharepoint-migration-assessment-toolscan-reports-roadmap` for insight and recommendations particular to your migration.

Next, we will review planning for content that cannot be migrated.

Determining the course of action for content that cannot be migrated

SPMT can migrate data and features from file shares or SharePoint sites to SharePoint Online sites, OneDrive for Business sites, or Microsoft Teams. In planning your migration, you should have determined the target migration locations for your data sources.

After running the migration assessment tool, you may have received warnings or errors that impact your migration. Some warnings may be able to be ignored (but may result in items listed not being migrated), but error items indicate features that will not be migrated or cannot be accessed.

When encountering issues regarding applications or services in the on-premises SharePoint environment, it's important to review the business requirements for both the existing and future environments and use that to inform what the best course of action will be. Potential courses of action include the following:

- Updating the source application or content to something that is compatible with SharePoint Online prior to migration.
- If the destination application or content *is* supported in SharePoint Online, the migration tool may not be able to process it. You may be able to use a third-party product to perform the migration.
- Deprecate the use of the application, content, or feature, and proceed.
- Leave the content or application, and continue to use it in its current state.

If it is determined that content cannot be moved, you'll need to revisit your migration plan and potentially re-evaluate the usage of any SharePoint Hybrid features to provide users with a functional experience.

Next, we'll look at configuring and executing a migration.

Planning and configuring an automated SharePoint migration

Now that you've identified and resolved issues prior to migration, it's time to migrate data. In these examples, we're going to refer to the native tools provided by Microsoft—namely, SMAT and SPMT.

Use the following steps to complete a migration:

1. On SharePoint Server with SPMT installed, launch the tool.
2. Sign in, if prompted, to your SharePoint Online tenant.
3. Click the **Start your first migration** button, as illustrated in the following screenshot:

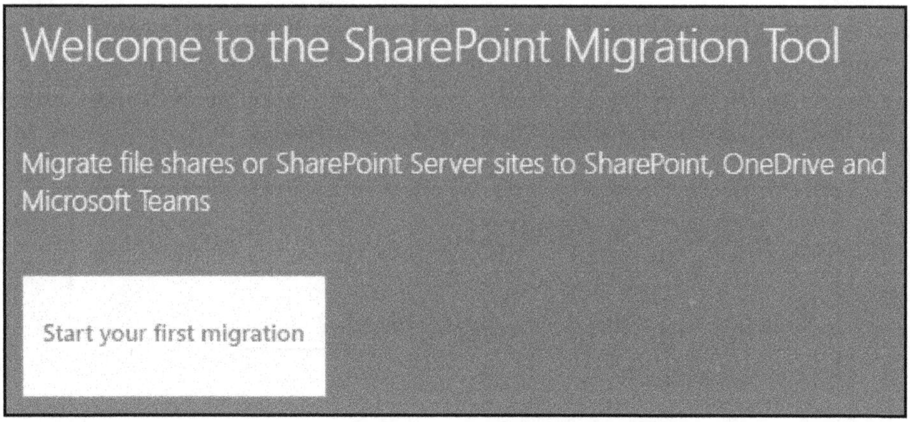

Welcome to the SharePoint Migration Tool

Migrate file shares or SharePoint Server sites to SharePoint, OneDrive and Microsoft Teams

Start your first migration

4. Select the data source. In this case, we'll choose **File Share**, as illustrated in the following screenshot:

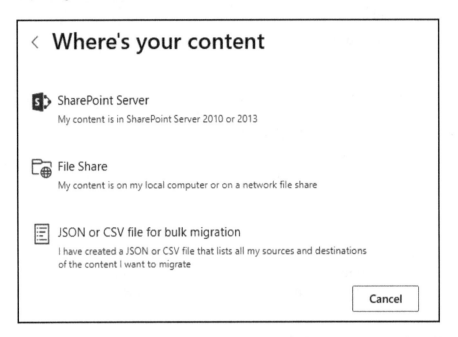

5. If selecting a SharePoint site, enter the **Uniform Resource Locator** (URL) for a SharePoint site and click **Next**. If prompted, enter local site credentials and click **Sign in**. If selecting a file share, enter the **Universal Naming Convention** (UNC) path to the share and click **Next**, as illustrated in the following screenshot:

6. Select a destination (**Microsoft Teams**, **SharePoint team site**, or a **OneDrive** (for Business) site), as illustrated in the following screenshot:

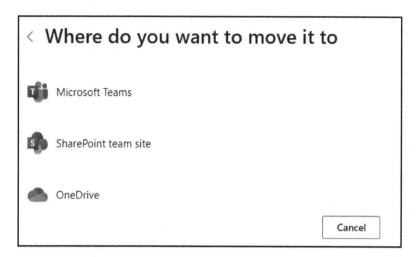

7. Select or input the destination. When selecting **Microsoft Teams** as the destination, the team *must already exist*. SharePoint sites can be created automatically. If using **OneDrive** as the destination, enter the user's email address. Click **Next**, as illustrated in the following screenshot:

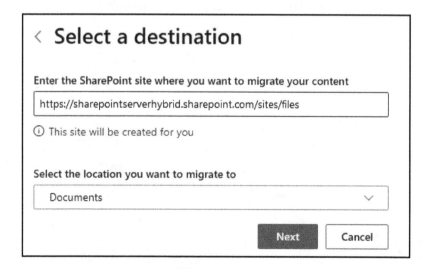

8. If you have more data to migrate, you can select **+Add another source** to restart the wizard at *step 4*. Continue until all additional sources and destinations are mapped, following the same process. When finished, enter a name for the migration and click **Next**, as illustrated in the following screenshot:

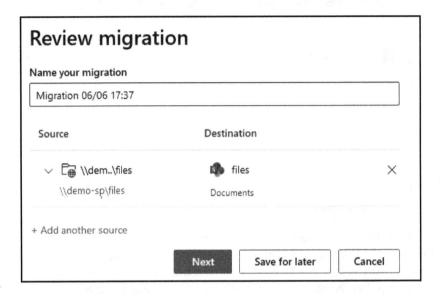

9. Click **Migrate**, as illustrated in the following screenshot:

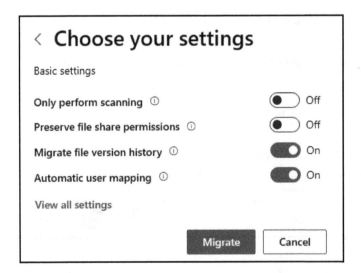

It's important to note in this step that you do have the option of just running a scan (sometimes referred to as a pre-flight or pre-migration test) to detect any issues during migration. If you select the **Only perform scanning** option, no data will be migrated, but you will get a report detailing the tool's findings. To do that, follow the steps given next:

1. Review the **Migration details** window for any issues or notifications, as illustrated in the following screenshot:

2. When the migration is finished, you'll receive a status output screen. Whether you did a full data migration or only a content scan, you'll have a **View reports** button available that will highlight any issues that were encountered, as illustrated in the following screenshot:

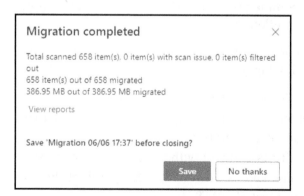

When the migration is finished, you should be able to navigate to the new location and view data.

Next, we'll look at some other recommended tools and strategies.

Recommended tools and strategies to migrate data

As discussed in the previous chapter, Microsoft provides several tools that can assist in planning for migrations or actually migrating data, including the following:

- **SMAT**: Planning and assessment for on-premises-to-cloud migrations
- **SPMT**: Content scanning and migration tool for on-premises-to-cloud migrations of SharePoint and file server content
- **SharePoint Migration Manager**: Migration tool for on-premises file shares to SharePoint Online
- **Mover.io**: Cloud-to-cloud migration scenarios, including SharePoint- and Google-based solutions

Each of those tools has its own use cases. For on-premises-to-cloud migrations, most organizations choose to use SMAT and SPMT—they have the longest track record, and both Microsoft support and the technical community are familiar with their configurations and outputs. SharePoint Migration Manager is a newer tool that only works against file shares at the moment, and Mover.io was acquired through purchase and is used for cloud-to-cloud migration scenarios primarily.

Microsoft currently recommends using SMAT and SPMT to migrate from on-premises SharePoint farms to SharePoint Online.

From a migration cadence and workflow perspective, it may be desirable to go through a process such as the following one:

1. Enable scripting support in SharePoint Online.
2. Identify content (sites, files, apps, lists, and libraries) that need to be migrated.
3. Download and extract SMAT.
4. Run an identity assessment with SMAT, and resolve issues.
5. Run a SharePoint assessment with SMAT, and resolve issues.
6. Download and install SPMT.

7. Add the source resources you wish to migrate and select the **Only perform scanning** option on the **Choose your settings** page of the migration tool.

8. Review the report and resolve any issues. Repeat until issues are resolved.

9. Develop a change management communication plan with business stakeholders.

10. Notify users that they will not be able to update any files while the migration is in process.

11. Set any file shares to **Read Only** to prevent users from updating files.

12. Complete or stop any workflows to prevent changes in the state due to workflows processing items.

13. Set SharePoint sites to **Read Only** (using `Set-SPSite -Lockstate ReadOnly`) to prevent users from updating files.

14. Run SPMT and migrate data to SharePoint Online.

15. Verify migration success. This process should involve technical staff to verify that the data was moved without any resulting errors (or report on any error conditions), as well as business owners, to ensure completeness and functionality.

16. Notify users of new file locations.

After your migration has completed, you'll want to ensure that users are accessing data and applications in the new Microsoft 365 environment. It may be advisable to leave the source content set to **Read Only** to prevent bifurcation of work and users updating data in two different locations.

Summary

In this chapter, we reviewed some additional planning and workflow around the actual migration process, including running SMAT and SPMT to ensure your environments are ready for migration.

SMAT provides a number of reports for readiness, including an identity report that will help you ensure that users have access to data post-migration.

You learned about the steps necessary to execute pre-flight and actual migrations, ensuring that you'll be able to successfully migrate the first time.

Finally, you learned some best practice processes for planning and executing migrations, including preventing changes to data and executing a change management communication plan.

Other Books You May Enjoy

If you enjoyed this book, you may be interested in these other books by Packt:

Workflow Automation with Microsoft Power Automate
Aaron Guilmette

ISBN: 978-1-83921-379-3

- Get to grips with the building blocks of Power Automate, its services, and core capabilities
- Explore connectors in Power Automate to automate email workflows
- Discover how to create a flow for copying files between two cloud services
- Understand the business process, connectors, and actions for creating approval flows
- Use flows to save responses submitted to a database through Microsoft Forms
- Find out how to integrate Power Automate with Microsoft Teams

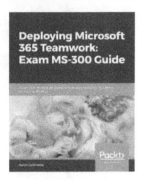

Deploying Microsoft 365 Teamwork: Exam MS-300 Guide
Aaron Guilmette

ISBN: 978-1-83898-773-2

- Discover the different Microsoft services and features that make up Office 365
- Configure cloud services for your environment and extend your infrastructure's capabilities
- Understand site architecture, site settings, and hub settings in SharePoint Online
- Explore business connectivity services for view and access options in SharePoint Online
- Configure Yammer to integrate with Office 365 groups, SharePoint, and Teams
- Deploy SharePoint Online, OneDrive for Business, and Microsoft Teams successfully, including bots and connectors

Leave a review - let other readers know what you think

Please share your thoughts on this book with others by leaving a review on the site that you bought it from. If you purchased the book from Amazon, please leave us an honest review on this book's Amazon page. This is vital so that other potential readers can see and use your unbiased opinion to make purchasing decisions, we can understand what our customers think about our products, and our authors can see your feedback on the title that they have worked with Packt to create. It will only take a few minutes of your time, but is valuable to other potential customers, our authors, and Packt. Thank you!

Index

about 178
process 179

G

granular backup
  about 44
  performing 45
group manager 250
groups 237, 247

H

high availability and disaster recovery options
  reference link 34
high-performance farm
  configuring 61
  core requisites 62
  network 71
  planning 61
  processors 63
  server infrastructure 62
  server roles 63
  storage requirements, determining 63, 64
host-named site collection
  creating 90, 91, 92
hub site architecture 478
hybrid app launcher
  configuring 376, 377, 378, 379
hybrid B2B sites
  configuring 371, 372, 374, 376
  using, advantages 372
hybrid configuration issues 386
  troubleshooting 383, 384, 385, 386
Hybrid Configuration Wizard
  about 391
  reference link 349
  used, for setting up cloud SSA 391, 392, 394,
    395, 396, 397, 398
hybrid federated search 389
hybrid OneDrive for Business, configuration
  about 363
  Hybrid Configuration Wizard, running 366, 368,
    369
  permissions, configuring 365, 366
  pilot group, creating 364, 365
  prerequisites 364

Hybrid Picker 391
hybrid scenarios
  evaluating 340
hybrid search topologies
  cloud hybrid search 389
  hybrid federated search 389
hybrid service integration
  planning 344, 345
Hybrid Sites
  App Launcher 341
  configuring 369, 370, 371
  features 340
  OneDrive for Business 341
  self-service site creation 341
  site following 341
Hybrid taxonomy 343
hybrid taxonomy and content types, configuration
  about 358
  Hybrid Configuration Wizard, running 361, 362,
    363
  on-premises taxonomy, copying to SharePoint
    Online 359, 360
  prerequisites 358
  term store permissions, updating 358

I

I/O Operations Per Second (IOPS) 63
Information Rights Management (IRM)
  planning for 48, 49, 50
Internet Information Services (IIS) 43, 86
Internet Protocol security (IPsec) 43
Internet Server Application Programming Interface
    (ISAPI) 43
Intrusion Detection (IDS) 479
Intrusion Prevention (IPS) 479

K

Key Performance Indicators (KPIs) 11, 482
Kutso Query Language (KQL) 305

L

language packs, for SharePoint Server 2016
  download link 271
language packs, for SharePoint Server 2019
  download link 271

www.ingramcontent.com/pod-product-compliance
Lightning Source LLC
Chambersburg PA
CBHW081452050326
40690CB00015B/2767